AF661325

ENCYCLOPAEDIA OF EMINENT THINKERS

VOLUME-XIX: **BHAGAT SINGH**

Volumes by the same author

The Political Thought of ***Dadabhai Naoroji***
The Political Thought of ***Tej Bahadur Sapru***
The Political Thought of ***C. Rajagopalachari***
The Political Thought of ***Subhas Chandra Bose***
The Political Thought of ***Bhagat Singh***
The Political Thought of ***Annie Besant***
The Political Thought of ***C.R. Das***
The Political Thought of ***Lalalajpat Rai***
The Political Thought of ***Sir Pherozeshah Mehta***
The Political Thought of ***Mahadev Govind Ranade***

About the Author

Dr. (Miss) Ashu Pasricha is a faculty in Gandhian studies at Panjab University, Chandigarh. She has many books to her credit including *Gandhian Approach to Integrated Rural Development; Public Administration: Discipline and Dimensions; Peace Studies: Discipline and Dimensions; WTO, Self-reliance and Globalisation; and Gandhi's Concept of State and Power.* Also, she has published a number of research papers in various Indian and foreign journals of repute.

ENCYCLOPAEDIA OF EMINENT THINKERS

VOLUME NINETEEN

POLITICAL THOUGHT OF BHAGAT SINGH

Dr. Ashu Pasricha
Department of Gandhian Studies
Panjab University, Chandigarh

CONCEPT PUBLISHING COMPANY, NEW DELHI-110059

Cataloging in Publication Data—DK
Courtesy: D.K. Agencies (P) Ltd. <docinfo@dkagencies.com>

Encyclopaedia of eminent thinkers / K.S. Bharathi.
v. 16-20 cm. — (Gandhian studies and peace research series ; 11)
Vols. 16, 19 by Ashu Pasricha; v. 17-18 & 20 by Jai Narain Sharma.
Includes bibliographical references.
Includes indexes.
ISBN 13: 9788170226840 (set)
ISBN 10: 8170226848 (set)

1. Statesmen—India. 2. Political science—India. 3. India—Politics and government. I. Bharathi, K. S. (Kuduva Swamy), 1956- II. Pasricha, Ashu, 1967- III. Sharma, Jai Narain, 1951- IV. Series: Gandhian studies and peace research series ; 11.

DDC 954.035 22

All rights reserved. No part of this work may be reproduced, stored in a retrieval system, or transmitted in any form or by any means, electronic, mechanical, photocopying, microfilming, recording or otherwise, without the prior written permission of the copyright owner and the publisher.

ISBN -13: 978-81-7022-684-0 (Series)
ISBN -10: 81-7022-684-8 (Series)
ISBN -13: 978-81-8069-499-8 (Vol. XIX)
ISBN -10: 81-8069-499-2 (Vol. XIX)

First Published 2008

© Concept Publishing Company

Published and Printed by
Ashok Kumar Mittal
Concept Publishing Company
A/15-16, Commercial Block, Mohan Garden
New Delhi-110059 (India)
Phones: 25351460, 25351794
Fax: 091-11-25357103
E-mail: publishing@conceptpub.com

DEDICATED
TO
ADVOCATE S.C. VERMA
AND
DR. RAJ VERMA

PREFACE

The revolutionaries were active in Indian politics for several years. They contributed a great deal in their own way towards the freedom of the country. Although they could not penctrate deeply in the hearts of the masses, they certainly infused in them a sense of patriotism and a determination to drive out the foreigners from their soil. This sort of spirit alarmed the British bureaucrats ruling the country at that time. Besides, they were also distressed to see that even those, who were opposed to their ideology and methods, praised them for their love of motherland and the heroic way in which they faced the gallows or an extremely hard life in the British jails of India including 'Kala Pani'. While denouncing their cult of violence even Mahatma Gandhi, an apostle of non-violence, unhesitatingly appreciated their feelings of intense patriotism and their willingness to sacrifice their all for the emancipation of their country from foreign yoke.

Undoubtedly the revolutionaries, except a few, treaded the thorny path, they had willingly chosen, with great courage and bravery. Several of them faced the gallows smilingly with great fortitude. They certainly deserve to be ranked among martyrs of freedom. Among such martyrs the name of Bhagat Singh shines as a star. He is rightly called the 'Prince of Martyrs'.

Bhagat Singh stated the gospel truth when he said, "You can kill individuals, but not the ideas. Great empires crumbled, but the ideas survived." The Bourbons, the Czars and the British Empire in India fell, but India as free Bharat survives all vicissitudes of foreign conquest and subjection. Indeed, this is what Bhagat Singh fought and died for. He wanted India not only to be free but also a sovereign socialist republic of workers and peasants. In the leaflet thrown in the assembly, he declared, "we dream of a glorious future, when man will be enjoying perfect peace and full liberty. But, the sacrifice of a few individuals at the altar of the great revolution that will bring freedom to all, rendering the exploitation of man by man impossible, is inevitable."

Bhagat Singh, symbolizing the revolutionary upsurge of his time, challenged the might of the world's greatest imperialist power. Commenting on his work, Netaji Subhas Chandra Bose said, "Bhagat Singh is today not a person, but a symbol. He symbolizes the spirit of revolt that has taken possession of the country... Bhagat Singh, who has set an example of character and patriotism by sacrificing himself for the sake of the country's freedom was from the community. Today, he is known to be a brave Sikh hero throughout the world. The Sikh community has to produce thousands of Bhagat Singhs for the cause of the country."

India's freedom movement constitutes a glorious chapter in the political history of India. Many strands went into its making, of which the revolutionary patriots were one. Bhagat Singh was perhaps the top ranking fighter of the revolutionary camp, a saga of struggle and sacrifice in the cause of India's independence.

A special feature of this book is the analysis of the political ideology of Bhagat Singh. To rate Bhagat Singh merely a freedom fighter is not enough. He was much more than that. He believed in and advocated scientific socialism. As he stated in the Delhi Court, "Labour is the real sustainer of society. The sovereignty of the people is the ultimate destiny of the workers. For the ideals and for this faith we shall welcome any suffering to which we may be condemned. To the altar of this revolution, we have brought our youth as incense, for no sacrifice is too great for so magnificent use. We are content to await the advent of revolution." While explaining the meaning of revolution in the court, he further said, "We mean by revolution the uprooting of the present social order—a change in the present order of things which is based on manifest injustice."

This pioneer attempt of presenting the political philosophy of Bhagat Singh will, it is hoped, elicit interest of the students of political science and history as well as the general reader.

Personal acknowledgments are no easy matters where so many have been kind. My special thanks are due to Professor R.N. Pal, Professor N. Radhakrishnan, Dr. K.S. Bharathi, Dr. A.D. Mishra, Professor Jai Narain Sharma, Dr. Meena Dutta, Dr. Seema Pasricha, Dr. Rashmi Sharma, Mr. Ajay Pasricha (advocate) and Master Hemant Malhotra. Beyond that I can make only an all embracing salute to the galaxy of the scholars whose works I have consulted.

I am also grateful to Shri Ashok Kumar Mittal, my publisher, for his consistent zeal in the printing and publication of this volume.

Chandigarh **ASHU PASRICHA**

Encyclopaedia of Eminent Thinkers Series

Vol. I. The Political Thought of **Mahatma Gandhi**/ *K.S. Bharti*
Vol. II. The Political Thought of **Nehru**/ *K.S. Bharti*
Vol. III. The Political Thought of **Rabindra Nath Tagore**/ *K.S. Bharti*
Vol. IV. The Political Thought of **Radhakrishnan**/ *K.S. Bharti*
Vol. V. The Political Thought of **Vinoba**/ *K.S. Bharti*
Vol. VI. The Political Thought of **Aurobindo**/ *K.S. Bharti*
Vol. VII. The Political Thought of **Jayaprakash Narayan**/ *K.S. Bharti*
Vol. VIII. The Political Thought of **Vivekananda**/ *K.S. Bharti*
Vol. IX. The Political Thought of **Ambedkar**/ *K.S. Bharti*
Vol. X. The Political Thought of **M.N. Roy**/ *K.S. Bharti*
Vol. XI. The Political Thought of **Dadabhai Naoroji**/ *Ashu Pasricha*
Vol. XII. The Political Thought of **Veer Savarkar**/ *Jai Narain Sharma*
Vol. XIII. The Political Thought of **M.A. Jinnah**/ *Jai Narain Sharma*
Vol. XIV. The Political Thought of **Tej Bahadur Sapru**/ *Ashu Pasricha*
Vol. XV. The Political Thought of **C. Rajagopalachari**/ *Ashu Pasricha*
Vol. XVI. The Political Thought of **Subhas Chandra Bose**/ *Ashu Pasricha*
Vol. XVII. The Political Thought of **Mohammad Iqbal**/ *Jai Narain Sharma*
Vol. XVIII. The Political Thought of **K.B. Hedgewar**/ *Jai Narain Sharma*
Vol. XIX. The Political Thought of **Bhagat Singh**/ *Ashu Pasricha*
Vol. XX. The Political Thought of **M.S. Golwalkar**/ *Jai Narain Sharma*

CONTENTS

1

HIS LIFE AND TIMES

Bhagat Singh was born on September 27, 1907 at Banga.[1] Banga is a village in Layallpura district of Punjab province. A brave man, Sardar Kishan Singh by name, lived there. Vidyavati was his wife. Kishan Singh's family was noted for bravery and love of adventure. Many heroes of his family had fought to free India from the British. Such fighters are called revolutionaries. Kishan Singh too was a revolutionary. His younger brothers, Ajit Singh and Swaran Singh, too, had fought to drive the British out of India. Kishan Singh, Ajit Singh and Swaran Singh had all been sent to prison by the Government. In those days such a revolution had spread all over the country. People were full of determination to win freedom. Pare the day Bhagat Singh was born, his father Kishan Singh was released on bail from the Central Jail of Lahore. His elder uncle Ajit Singh was released on the same day from the Mandalay Jail. His younger uncle Swaran Singh was also released from the Central Jail of Lahore. The release of all the three brothers from different jails on the day of the birth of Bhagat Singh was considered very auspicious for the family. Therefore, the grandmother of Bhagat Singh initially named the new born child as Bhaagan Wala. Later on the child was named as Bhagat Singh.[2] 'Bhagat Singh' means 'the fortunate'.

Bhagat was a lovely child. His smile was charming. People used to say that he would become very famous. His mother Vidyavati's life had been full of sorrow right from the beginning. The revolutionary husband would always be away. Always lurking in Vidyavati's mind was the fear that he might at any time be sent to jail. It was a family of fighters for freedom and one or the other would always be in jail. Vidyavati herself had to look after the affairs of the family. At such anxious times, her children were her only comfort. They were intelligent and brave and this made her forget her misery. Bhagat Singh was her favourite.

Bhagat Singh was admitted to the primary school in Banga. From his childhood he was highly interested in studies and was ahead of the others in his class. He used to write a beautiful hand. He was the favourite pupil of his teachers. Very much liked by his classmates, he was their leader. Big boys used to carry Bhagat Singh on their shoulders to the school and back home. His childhood itself indicated that later he would become a leader of revolutionaries.

Bhagat Singh easily made friends with one and all. His companions were naturally his friends. But cartmen and coolies, and the very men who swept the streets were also his friends.

Once clothes had to be stitched for Bhagat Singh. The old tailor who stitched clothes delivered them at the house and went away. "Who is that who brought the clothes?" asked mother, Vidyavati.

> "My friend," replied Bhagat Singh.
> "What! Is the tailor, too, your friend?",
> Vidyavati asked, surprised.
> "Yes, everyone in the village is my friend", was Bhagat Singh's reply.

Thus the ability to win the hearts of men grew in Bhagat Singh right from his childhood.

Bhagat Singh had two uncles. Of them, Swaran Singh had again been sent to prison by the British. Life in prison was wretched and he fell ill. His health did not improve even after his release and he died. When Ajit Singh came out of the jail, he left the country. Bhagat Singh's aunts would often recall their husbands' misery, and lament over it. Seeing this, Bhagat Singh would bravely say, "Don't weep, aunt. When I grow up, I will drive out the British and bring back my uncle. I will take revenge upon the British who are the cause for my uncle's illness." On hearing the heroic words of the little boy, the weeping women would burst into laughter. At least for the moment they would forget their sorrow.

When he was in the fourth class, Bhagat Singh asked his classmates, "What do you wish to become when you grow up?"

Each boy gave a different answer. "I intend becoming a doctor," said one. Another said, "I will be a government officer." Still another would become a merchant; while another 'would marry'. Bhagat Singh remarked, "Is marriage a big achievement? Anybody can marry. On the other hand, I will drive the British out of India."

Thus patriotism flowed in his veins from the days of his childhood. By the time he completed his secondary education, Bhagat Singh knew everything about the revolutionaries of his family. He had read all the records about them at home. And the desire to fight for the country's freedom grew strong in him. Bhagat Singh finished his primary education at Banga.

From his childhood Bhagat Singh showed signs and traits of a great man. At the age of two and a half years he accompanied his father to the farm in the village where mango sapplings were being planted. In the farm when Kishan Singh was engaged in a deep conversation with his friend Mehta Anand Kishore, an enthusiastic worker of the Bharat Mata Society, Bhagat Singh broke some twigs and began to plant them in the ground. On seeing this, Mehta Anand Kishore remarked to Kishan Singh, 'your son is already a farmer and planting trees.' Kishan Singh now asked his son as to what he was doing. Bhagat Singh replied innocently, 'Babboke Bo Raha Hun' meaning that he was planting 'banduks', i.e. guns. Evidently, the child was not able to pronounce 'banduks' correctly.[3] This remark of Bhagat Singh can be regarded as a symbol of his future activities. It was also in consonance with the atmosphere prevailing in the family at that time. Freedom fighters frequently visited Banga to discuss political matters with Arjun Singh and Kishan Singh. Evidently Bhagat Singh must have overheard the discussions and as an intelligent child he must have pondered over such talks.[4] Besides, he was deeply impressed by the tales of sacrifice of Ajit Singh narrated to him by Harnam Kaur.[5] It created resentment in his mind against the atrocities of the British government. During these talks he would comfort his aunt, 'Chachiji (Aunt), I will not rest until I have driven the British out of the country.'

He was an intelligent, hardworking and a disciplined student. He was very popular with the teachers and the students. His teachers were satisfied with his performance. As a young boy he had studied the books written by Lala Hardayal, Sufi Amba Prasad and Ajit Singh on the revolutionary movements of the world, as these works were available in his home. Moreover in Banga, his grandfather Arjun Singh constantly inspired him.[6]

The 'Ghadar' rebellion during the First World War had profoundly affected Bhagat Singh. In the first decade of the twentieth century a sufficient number of Indians, mostly Punjabi Sikhs, had migrated to Canada and some areas in the United States of America on the Pacific Coast. Simultaneously several prominent Indian revolutionaries had migrated to San Francisco, Los Angeles and other towns on the East Coast of America. They hoped that they would get funds and recruits by appealing to the sentiments of patriotism of these settlers who were dissatisfied with the treatment given to them in Canada and the failure of the British government to protect their interests in a commonwealth country. Another important reason was the belief of the Indian revolutionaries about America being 'the land of freedom and opportunity.'[7] They also believed that this was the only country which could resist the pressure of the British government.

These self-exiled Indian revolutionaries established several associations in the major cities of America for anti-British propaganda. In a meeting held in San Francisco sometime at the end of 1912, a few Indians living in the States and Canada decided to establish an organization called 'The Hindu Association of the Pacific Coast.' Hardayal and Sohan Singh Bhakna were elected Secretary and President respectively of the newly formed association. In March 1913, the Association decided to publish a paper named '*Ghadr*' from San Fransisco. Its first issue was published on November 1, 1913. Through this paper an intense anti-British propaganda was launched. The paper was widely circulated in America, China, Japan and Hong Kong. Successful attempts were also made to circulate it widely in India. On account of the paper 'Ghadr' the association came to be known as Ghadar Party. The Sedition Committee reported that the Ghadr newspaper and its progeny were distributed in every place where the revolutionaries hoped to gain adherence, particularly among troops.[8] It appears that the issues of Ghadr, which stressed on the weakness of the British government and exhorted Indians in the mainland to co-operate in the efforts of freedom from the foreign domination, were also read by Arjun Singh and his family.

With the commencement of the First World War the activities of the Ghadr Party took the shape of a rebellion against the British government in India. The Komagata Maru incident, which though not directly a part of the activity of Ghadr Party, strengthened the hands of Indian revolutionaries abroad who were urging Sikhs living in the Pacific coast of Canada and United States of America to return to India and join the mutiny which they asserted was about to begin in their motherland.[9] Bhagat Singh was greatly perturbed over this incident. In April 1914 a Japanese ship named Kamagata Maru was charactered by Gurudit Singh. It left Hong Kong with Indian passengers from that city, Shangai, Moji and Yokabame ports for Vancouver port of Canada. At Vancouver port the Canadian government did not allow the Indians to land. A petition submitted by the 'Shore Committee' consisting of Indian immigrants living in Vancouver was rejected by the Supreme Court of Canada. During the return journey of the ship First World War broke out in Europe. The passengers of the ship were not allowed to disembark at Hong Kong and Singapur. On September 20, 1914 the ship reached Budge-Budge harbour near Calcutta on the river Hooghly. The local police asked the passengers to board a train directly going to Punjab. The passengers were roughly treated by the police. It resulted in a riot in which the troops standing nearby opened fire. Eighteen Sikhs were killed on the spot and twenty five were wounded. The incident was condemned by the press, specially the Sikh papers of Punjab. It helped in the popularity of the Ghadr Party.

The Ghadr leaders decided to organise a rebellion in India. From America they sent several revolutionaries to India. The main centre of their activities was Punjab. Majority of the Ghadr volunteers reached Punjab in or about November, 1914. Arjun Singh's family was very keen to know about the progress of the Ghadrites. Their conversations about the Ghadar movement, Lala Hardayal, Sohan Singh Bhakna, Gurdit Singh, Ras Behari Bose and Kartar Singh Sarabha were heard attentively by Bhagat Singh. In February 1915, Kartar Singh Sarabha, who was financially helped by Kishan Singh, was arrested at Shahpur. He was sentenced to death in the first Lahore Conspiracy case. The execution of Kartar Singh Sarabha at the age of nineteen profoundly affected Bhagat Singh. The life and sacrifices of Sarabha seems to have surpassed all other influences on the political philosophy of Bhagat Singh. The extent of impact of martyr Kartar Singh Sarabha's heroism and sacrifice on Bhagat Singh can be judged from the fact that at the time of his arrest in 1929 a photograph of Kartar Singh Sarabha was recovered from him. It appears that he always carried the photograph with him. Moreover, he used to show the photograph of Sarabha to his mother and say, "Dear mother, this is my hero, friend and companion." While at home, Bhagat Singh used to sing the following couplet which was sung by Sarabha. The couplet runs as follows:

> "Seva desh di jindarhiye barhi aukhi
> Gallan kamian dher sukhalian ne
> Jinon desh seva vich paer paya,
> Onan lakh museebatan jhalian ne."[10]

Ras Bihari Bose and other leaders of the Ghadr movement as well as the leaders of anti-British movement in Punjab had much impact on Bhagat Singh. Some of the leaders of Punjab occasionally visited Banga to seek the advice and financial support of Kishan Singh.

In 1916-17, after finishing his primary education at Banga, Bhagat Singh joined the Dayanand Anglo Vedic Middle School, Mohan Lal Road in Lahore. By this time his father Kishan Singh had shifted to Nehrankot area in Lahore as he was working as an agent in the Prudential Insurance Company. The school was soon raised to High School status. In those days this school was regarded as a premier academic institution in northern India as it produced a rich crop of patriots and bureaucrats. All the teachers of this school were highly qualified, well trained and above all sincerely devoted to their profession. As such the institution attracted the best students belonging both to the elite and middle class families. Bhagat Singh's performance in school was quite satisfactory. A postcard written in Urdu by Bhagat Singh on July 22, 1918, when he was only eleven years old, to his grandfather Arjun Singh is a testimony to it. In this letter he informed his grandfather that in sixth class he had secured 110 marks out of 150 in Sanskrit and 68 out of 150 in English. He hoped that he would secure good marks in other subjects also.[11]

Along with regular studies Bhagat Singh devoted considerable time in studying literature relating to the exploits of several patriots and nationalists of India and other countries. His mother, who was always concerned about his following the foot-steps of his father or more so of his uncle, constantly reminded him that he has to look after the family. She did not like the types of books he read. She knew for certain that they were not part of the school curriculum. She was even more worried when he took notes from them with reverence.[12] Besides, he keenly watched the actions of the British government in India and the reactions of the Indian leaders to such events.

Meanwhile, extremists came to the forefront in the Congress. Under the leadership of Bal, Pal and Lal, the extremists launched an agitation against the injustices of the British government. But they did not believe in violence. However, the militant wing of the extremists, known as revolutionaries, subscribed to violence. For the sake of freedom of their country they tried to create terror among the British bureaucrats and their loyal supporters. During the initial years of the war the Ghadr rebellion was a serious threat to the British government, specially in the province of Punjab.

On account of several other reasons, signs of rising discontent were also noticed in Punjab. The indiscriminate recruitment of soldiers in the army, the war time exactions and the constant condemnation of the educated people of Punjab made its government unpopular.[13] Michael O'Dwyer, the Lieutenant Governor of the province, believed in suppressing all movements in his province with a strong hand. The end of the First World War brought in its wake serious problems on the economic and political front. The rise in prices, specially of foodgrains, affected the people adversely. There were *hartals* by labourers in several cloth mills of Bombay and other cities. The unemployment problem became more acute.

Thousands of people who were engaged in manufacturing articles required for the war lost their jobs. In Punjab, a large number of soldiers, who were discharged from the army, were seeking employment. Due to stern measures taken by Michael O'Dwyer, the Lieutenant Governor of Punjab, the situation was getting tense day by day.

During the war the Government of India had enacted Defence of Indian Rules. As the period for which it was enforced was to end shortly, the government wanted to maintain it in some form or the other even after the war. In 1918, a Sedition Committee was appointed by the Government of India to suggest measures for the suppression of revolutionary activities. Justice Rowlatt was appointed its chairman. In February, 1919 on the basis of the recommendations of the Sedition Committee, the Government of India introduced two bills in the Imperial Legislative Council. According to the two bills any person, who was suspected, could be imprisoned and without a trial he/she could be kept in jail for an indefinite period or punished after a secret trial.[14] In spite of remonstrances by the political parties and the people, these bills were passed in March 21, 1919. These bills are known as Rowlatt Acts. The passage and enactment of the Rowlatt Acts created a strong sense of resentment among the people. Gandhiji declared that he would offer *satyagraha* in case these acts were not withdrawn. But the government did not pay any heed to such remonstrances.

The patriotic fervour and peaceful resentment shown by the people irrespective of their affiliations to any caste, creed or community[15] was unprecedented. Michael O'Dwyer and his British subordinates were virtually scared of the Hindu Muslim Sikh unity expressed in the demonstration against the acts. The *hartals* observed in Amritsar on March 30 and April 6, 1919 were peaceful. The Hunter Committee appointed to investigate the Jallianwala Bagh tragedy later on reported that Muslims had taken part in a peaceful procession in large numbers which was taken out in Amritsar on April 9, 1919 on the occasion of Ram Navmi. But at some other places there were cases of mob fury causing damage to government buildings and property.

On April 9, 1919, Satyapal and Kitchlew, the two prominent leaders of Amritsar, were arrested. On the next day the Punjab government further precipitated the situation by firing on a peaceful crowd of demonstrators in Amritsar. The crowd turned into a mob and attacked post offices, banks, railway station and town hall. Five Englishmen were killed by the mob. The Government of Punjab sent General Dyer to Amritsar to control the situation. He reached Amritsar on April 10, 1919. On the next day martial law was declared by the government in Punjab. It transferred civil power to military authorities.

Besides Amritsar, there were disturbances in Lahore and several other towns of Punjab. The five districts affected seriously were Amritsar, Lahore, Gujranwala, Gujarat and Lyallpur.

On April 13, 1919, the day of Baishakhi, a meeting was held at Jallianwala Bagh in Amritsar. Although General Dyer had declared it unlawful, majority of the people attending it were unaware of it. Without giving any warning Dyer and his soldiers fired indiscriminately on the people present in the meeting. According to reports of the government, four hundred persons were killed and about two thousand were injured. But the number of those killed as well as injured was much more than what was reported by the government.

The Jallianwala Bagh tragedy shook the entire nation. Bhagat Singh was then twelve years old; his mind was deeply disturbed by this event. The next day he did not return home after the school hours. His people at home waited and grew anxious.

Instead of going to school, Bhagat Singh went straight to the place of the tragedy. Somehow managing to push through the police on guard, he went in. He collected a bottle of mud wet with the blood of Indians and returned home. Seeing that he was late, his younger sister said, "Where were you all this time? Mother has been waiting to give you something to eat." But Bhagat Singh was not at all thinking of food. Showing the bottle in his hand, he said, "Look here. This is the blood of our people killed by the British. Salute this."

Then he put the bottle in a niche and worshipped it with flowers. The people who had assembled in Jallianwala Bagh carried no weapons. Nor was there a way to escape from the place. And these people were killed by the British bullets! Such were the thoughts working in the mind of Bhagat Singh. The feeling that somehow the British must be driven out of India became firmer.

"That was the time when the Indian National Congress was fighting for the country's freedom. It awakened the people's love for their country and was uniting the people. Even before entering the ninth class, Bhagat Singh decided to take up this work. He was only thirteen. Bhagat Singh told his father of his decision and asked for his permission. Himself a revolutionary, Kishan Singh willingly gave his consent. Bhagat Singh left the school and joined the movement.

At that time, there was a powerful anti-foreign cloth movement in the country. If foreign cloth is bought, other countries are benefited. In order to end this, we have to wear cloth made in our country. Foreign cloth must be burnt—so the leaders taught. Bhagat Singh took part in this movement with zeal. Right from his early days, he used to wear only Khadi. With what zeal he helped the cause of homemade cloth and burnt foreign cloth! Every week he would collect foreign clothes, heap them up and burn them.

In 1920, Sardar Kishan Singh shifted his residence from Nawan Kot. According to Jaidev Gupta, a class fellow of Bhagat Singh, after the tragedy of the Jallianwala Bagh, the atmosphere in the house of Kishan Singh became very tense. All types of people used to come there and talked about martial law and its effects on the people of Punjab. Among these persons Bhai Parmanand, Kedar Nath Sehgal and Lala Pindi Das were prominent.[16] Now and then Bhagat Singh used to overhear their conversations.

Even after the tragic event of Jallianwala Bagh, the people in Amritsar, Lahore, Gujranwala, Kasur, Gujarat, Lyallpur and other areas were subjected to inhuman treatment by the martial law authorities. The crawling orders, flogging of innocent persons in public, arrest and convictions of old and infirm persons, and people of all walks of life without affording them any kind of legal assistance to plead their innocence; unethical interference in the functioning of numerous educational institutions, including the prestigious Medical College, Lahore and infliction of punishments on the students in the hottest days of May 1919 without the permission and assent of the educational authorities, indiscriminate firing at numerous places in the name of law and order resulting into the deaths of many innocent persons, the posting of punitive police; the exactions of illegal fines on the areas as a whole procuring numerous articles like milk, ghee, egg, fruit, vegetables etc. by the army personnel without any kind of payment to the shopkeepers, compelling innocent people to give false evidence to support government's versions and the hanging of numerous persons who had participated at one time or another in protests had serious impact on the minds of the people of Punjab. Bhagat Singh was also very much distressed about hearing such happenings. In such a tense atmosphere in December 1919, the Indian National Congress held its sessions at Amritsar under the presidentship of Moti Lal Nehru, who criticised the British government in many ways. His speech enthused the people of India with new dimensions of political ideology.[17]

In May 1920, the report of Hunter Committee on the events in Punjab during 1919 was published. According to this report general Dwyer was not guilty. Along with it the statement of General Dyer before the Hunter Committee that he wanted to teach the people a lesson by firing on them and that his aim was not only to disburse the crowd in Jallianwala Bagh but was also to produce a 'moral effect', created resentment among the Indians. The refusal of the House of Lords to pass a motion to condemn General Dyer was also denounced by them. The Morning Post, a newspaper of England, had collected an amount of £ 26,000 for presenting it to Dyer. All these decisions greatly hurt the feelings of whole of the nation.

In the special session of the Congress held at Calcutta from 4 to 9 September 1920, a programme to renounce the honours and titles awarded by the British government; boycott of educational institutions, courts and councils; boycott of foreign goods and promotion of 'khadi' was adopted. Bhagat Singh was highly impressed by this programme, specially that of opening of a chain of 'Vidyapiths', i.e. national schools and colleges in different parts of the country. In such institutions much emphasis was to be laid on the study of history, culture, traditions, philosophy etc. with Indian bias.

In 1922, the Congress organized a procession in the town of Chauri Chaura in Gorakhpur District. Then some rogues locked up twenty-two policemen together in a house, set fire to the house and burnt them. Before this, similar acts of violence had occurred even in Bombay and Madras. Mahatma Gandhi felt very sad at all this. He asked the people to end the non-cooperation movement which was then going on in the country.

That was a great disappointment to young Bhagat Singh, a lad of fifteen. Should an important movement be given up, just because 22 persons died? Before that, a nineteen-year-old revolutionary by the name Kartar Singh had been hanged by the British Government. Then none of these supporters of non-violence raised any objection. How could non-violence become so important now? Such thoughts weakened Bhagat Singh's faith in non-violence and non-cooperation movements. He went on firmly believing that armed revolution was the only practical way of winning freedom.

He made a deep study of the lives of the revolutionaries of Ireland, Italy and Russia. The more he read, the deeper grew his belief that war alone could bring freedom. The youth of the land should be inspired to turn to revolution. The very thought of fighting for freedom should thrill the young men. So thinking, Bhagat Singh began to organize the youth.

In August 1921, he left the D.A.V. School in ninth class and joined the non-cooperation movement. He contributed wholeheartedly in mobilising the people to offer their resistance to the British rule. After some time he decided to join National College, Lahore, which was founded and organised by Bhai Parmanand and Lala Lajpat Rai.[18] Although he was not eligible for admission in the college but in view of his intelligence and vast readings it was decided to consider his admission as a special case. Parmanand gave him two books for study. After two months a test was taken in which Bhagat Singh passed with credit and he was admitted in the college.[19] This college had been started by great patriots like Lala Lajpat Rai. Though he had not been to school for some years, Bhagat Singh had a good knowledge of history and politics. The Principal was astonished and permitted him to join the college straight away.

During the day he would listen to the lessons in the class. In the evening he would collect several friends and discuss the coming revolution. This became his daily routine. In an interview Chabil Das, a teacher of the National College, has described about the atmosphere prevailing in the college at that time. He mentioned that the teachers had personal relations with the students. In the evenings they met and talked informally about several subjects. They spent several moonlight nights on the banks of the river Ravi boating and rowing. During such excursions the students expressed their inner most feelings on any subject. Chabil Das further observed that there were no books at hand. The teachers selected books from the libraries and gave the relevant portions to the students. In the college the teachers and students talked about Ireland, about Sinn Fein movement of De Valera, other movements and the Russian revolution.[20]

At National College, Lahore, Bhagat Singh was highly impressed by Jai Chandra Vidya Alankar who taught history to the students. Vidya Alankar laid much stress on the study of the revolutions and social movements in different parts of the world. The success achieved by the Russian revolutionaries in 1917 against the Czarist regime was narrated by him to the students. He also acquainted them with the activities of the Indian revolutionaries. In this manner he inspired his students and tried to inculcate the spirit of sacrifice in them for the freedom of the country. The sacrifices of the Russian revolutionaries against the autocratic regime of the Czar impressed Bhagat Singh very much. He began to think about sacrificing his life for the freedom of his country from the shackles of the slavery.

In the college Bhagat Singh was regarded as an intelligent, active and smart student. He was well disciplined. He respected and obeyed his teachers. The teachers liked him very much. He was devoted to his studies. His academic results were very good. History and Political Science were his favourite subjects. Besides, he was a superb actor and a dramatist. A regular member of the local National Club, he often acted in the dramas and played the historical roles of the Mauryas and Guptas. He acted in a play named 'Chandragupta Maurya's Jug da Pahufutala'. In a play entitled Maharana Pratap, he acted as Pratap. All such plays had a contemporary reference to the freedom movement. These plays were performed with the active co operation of Progressive Writers and Actors of the National Drama Club of Lahore.

Bhagat Singh's immense success in the extra curricular activities, including the dramas, was mainly due to his typical Punjabi youthful physique and a commanding voice which always enabled him to dominate the stage by successfully performing the role of a hero in the drama or one act play.[21]

Bhagat Singh had a natural command over Punjabi, Urdu and English. As already stated he knew Hindi and Sanskrit very well. With all these qualities he gained much prominence amongst the teachers and students, not only of his college but of other local colleges also.[22]

While studying at National College, Lahore, Bhagat Singh maintained a close relation with his grandfather and other members of the family living at Banga. In a postcard of November 14, 1921 he enquired about the health of each and every member of the family. This letter also reveals that even at the age of fourteen Bhagat Singh was watching the political events with keen interest. In this letter he mentions that the employees of the railway are preparing for a strike and it is hoped that it would start in the next week. The letter also reveals his alertness in all matters. The news of the strike in railways was written in between the lines in opposite direction.[23]

An article written by Bhagat Singh in 1922-23 on the problem of Punjabi language and script shows his maturity in learning of different subjects.[24]

Some time in the end of 1923, Bhagat Singh discontinued his studies in Lahore. The immediate cause of this decision was the resolve of Bhagat Singh not to marry. His grandmother, who lived at Banga, was keen to perform his marriage soon. At that time he was about sixteen-year-old. Sardar Kishan Singh had contacted Sardar Teja Singh Mann of village Mannawala in Sheikhupura district for his sister's engagement with his son. Before finalizing the matter

Kishan Singh wrote to his son, "A decision has been made about your marriage and you will have to abide by the wishes of the family." Bhagat Singh consulted Chabil Das about it. In his talk with Chabil Das, he told that he was not prepared to marry. In reply Chabil Das said, "Bhagat Singh if I could get a really very good life companion who instead of retarding my activities, invigorates them, what would be your view ?" In the same breath Chabil Das quoted the concrete examples of Mrs. Sun-Yat-Sen, wife of Lenin and companion of Karl Marx. On hearing this Bhagat Singh said, "Guruji who can beat you in any argument ?"[25] Although it appeared to Chabil Das that Bhagat Singh had agreed to marry but this was not the case. In order to extricate himself from a difficult situation he decided to leave Punjab without informing any one. Before leaving Lahore he left a letter in the drawer of the table lying in his father's room. In this letter he wrote, "The mission of my life has been devoted to the attainment of freedom of motherland. Therefore, I do not give any kind of preference to rest and worldly desires." In the second para he wrote, "You must remember when I was a child, my grandfather announced at the time of my thread ceremony that my life would be dedicated for the cause of the motherland. I am, therefore fulfilling the commitment made at that time. I trust you would excuse me."[26]

In this connection Jaidev Gupta had a talk with Bhagat Singh. On being asked by Jaidev Gupta as to why he was not marrying, Bhagat Singh replied that he had chosen a path which was full of many possibilities. His two uncles had gone that way and they had left two widows. Should we leave another widow for weeping and wailing ? He told Jaidev Gupta that he was not going to spoil the life of a young lady. My determination is this and let these people say anything, do anything. The path he has chosen was fraught with serious consequences, even death. So marriage would be an obstacle in the way.[27]

It appears that even before the issue of marriage cropped up, Bhagat Singh was contemplating to leave studies for participating in the political movement. Jaidev Gupta has recorded an incident in his interview about their intention to do so. On one day Jaidev expressed his desire to Kishan Singh. On hearing this proposal Kishan

Singh was very angry. He remarked, "that would do, if you do not study '*Mazdoor Banoge Kya*', '*Dhakka Khaoege*', '*Pahele Parh to lo*." Jaidev then informed Kishan Singh that they would join National College. Similarly when they informed Lala Lajpat Rai about their intention to leave studies, he was also very angry. He advised both of them to continue their studies and become learned men. Only then they should think of joining the national movement.

From the letter left by Bhagat Singh in his home it is evident that he was seriously contemplating to tread on the path of revolution. He expressed his firm determination to work for the freedom of his country in his own way. In this connection Manmathnath Gupta has narrated an incident. According to him, on one evening Bhagat Singh and Yash Pal were jogging on the banks of Ravi. Suddenly, Yash Pal asked Bhagat Singh whether he would like to pledge his life for the country. At this moment both of them became sentimental and pledged their lives for the motherland.[28]

Thus, the reason for discontinuing studies and leaving Lahore was not simply an escape from the bonds of marriage. The real reason behind it was his desire to join the revolutionary movement. It appears that Jai Chandra Vidya Alankar had helped him in taking the final decision. Jai Chandra Vidya Alankar wrote a recommendatory letter addressed to Ganesh Singh Vidyarthi, who was doing secret work with great patriotic fervour in Kanpur. Vidyarthi was running a small press known as Pratap press. He was also publishing a paper named 'Pratap'. In Kanpur revolutionaries from Bengal, Bihar and United Provinces (Uttar Pradesh) used to come secretly and discussed their plans. It was therefore, considered proper by Vidya Alankar to send Bhagat Singh to Kanpur. Bhagat Singh reached Kanpur. First he earned his bread there by selling news papers. Then he came to know a revolutionary by the name Ganesh Shankar Vidyarthi. He got a job in the office of his periodical '*The Pratap*'. And he also learnt his first lessons as a revolutionary. Revolutionaries generally change their names so that people may not know them. Bhagat Singh became Balwant Singh. At home Bhagat Singh's parents were much worried about their son. His grandmother, then seriously ill, was very eager to see her grandson. His people searched for him and brought him back.

Genealogy of Bhagat Singh beginning with his grandfather Arjun Singh[29]

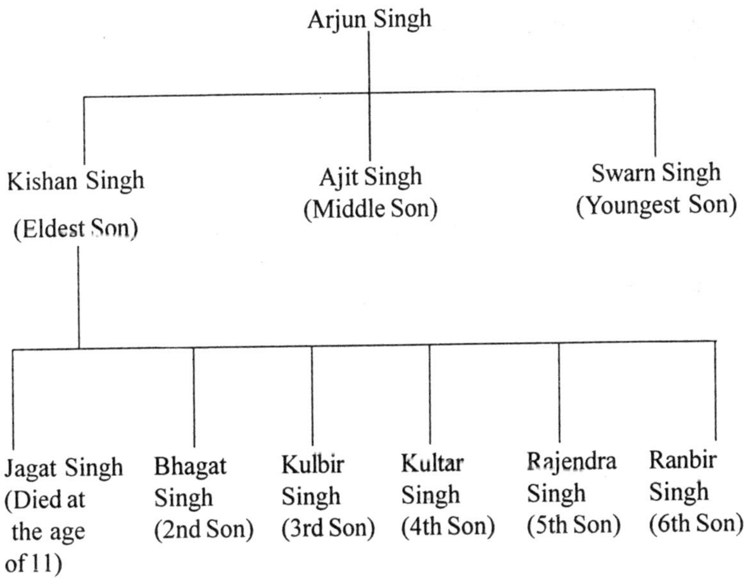

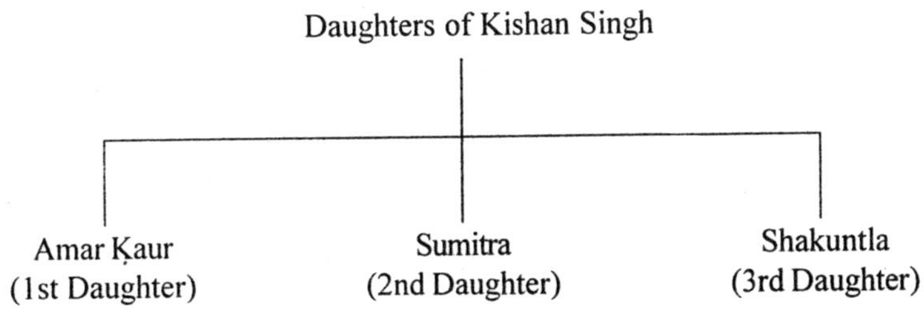

Note: All the daughters of Kishan Singh were younger to Bhagat Singh.

Even at home, Bhagat Singh could not sit idle. At that time the Akali Dal arranged a procession. But in order to prevent it, the District Collector Dil Bagh Singh issued an order: Nobody should supply either food or drink to the members of the Akali Dal.

The Collector who issued the order belonged to Bhagat Singh's family. But being a government officer, he hated the revolutionaries.

Bhagat Singh thought of helping the Akali Dal people visiting his village. He explained the situation to the villagers and arranged to supply food secretly to the Akali Dal people at night. Thus a week passed. The Dal's programme went on continuously and successfully. Throughout the day, there used to be talks on the country's freedom and the duty of the people.

Bhagat Singh also used to speak. The Collector grew angry that the people had helped the Dal against his orders. He issued a warrant to arrest Bhagat Singh. Then Bhagat Singh was only seventeen. So he was a minor and could not be arrested. The Collector grew angrier. "Bhagat Singh may be too young but his brain is not too young!" he grumbled.

It has already been mentioned that Bhagat Singh worked as a volunteer in the Non-cooperation Movement. However, he was disillusioned when in February, 1921 Gandhiji suddenly suspended it. Bhagat Singh felt that India had lost a great opportunity to bring the enemy to its knees. The loss of few days, even a single day, postponed not only the deliverance of the people but also defeated the psychological moment. In his view, the important point was to get rid of the rulers, not the manner in which it was done. He told his father that he did not either understand Gandhi's political strategy, nor his mental approach which had hacked the movement to pieces with a single blow. Although he had no faith in Gandhi's leadership or his creed of non-violence yet he praised Gandhiji for the immense awakening which he had generated through the Non-cooperation Movement. From this moment Bhagat Singh swinged towards the cult of revolutionaries. By the end of 1923, he had decided to follow the path of revolution.

REFERENCES

1. Deol, Gurudev Singh, *Shaheed-e-Azam Sardar Bhagat Singh* (Ludhiana: Deep Prakashan), 1978, p. 8.
2. Bakshi, S.R., *The Struggle for Independence* (New Delhi: Anmol Publications), 1989, p. 5.
3. Deol, Gurudev Singh, *Shaheed-e-Azam Sardar Bhagat Singh, op. cit.*, p. 6.
4. Singh, Swarn, *Path of Revolution: A Bibliography of Shaheed Bhagat Singh* (Dehli: Anmol Publications), 1990, p. 21.
5. Hamam Kaur was the wife of Ajit Singh.
6. Singh, Swarn *Path of Revolution: A Bibliography of Shaheed Bhagat Singh, op. cit.*, pp. 21-22.
7. Home, Political Branch, December, 1914, Number 96-98.
8. *Ibid.*, October, 1915, No. 91.
9. Rowlatt (Sedition) *Committee Report*, Chapter XI.
10. Singh, Swarn *Path of Revolution: A Bibliography of Shaheed Bhagat Singh, op. cit.*, p. 22.
11. Sandhu, Virendra, *Shaheed Bhagat Singh* (Delhi: Anmol Publications), 1974, p. 13.
12. Nayar, Kuldip, *The Martyr : Bhagat Singh Experiments in Revolution* (New Delhi: Har-Anand), 2000, p. 80.
13. *Ibid.*
14. Bakshi, S.R., *The Struggle for Independence, op. cit.*, p. 9.
15. Singh, Swarn, *Path of Revolution: A Bibliography of Shaheed Bhagat Singh, op. cit.*, p. 23.
16. *Transcript of an Interview with Jaidev Gupta*, Nehru Memorial Museum and Library, New Delhi.
17. Bakshi, S.R., *The Struggle for Independence, op. cit.*, pp. 9-10.
18. *Transcript of an Interview with Chabil Das*, Nehru Memorial Museum and Library, New Delhi.
19. Gupta, Manmath Nath, *Bhagat Singh and His Times* (Delhi: Lipi Prakashan), 1977, p. 95.
20. *Transcript of an Interview with Jaidev Gupta*, Nehru Memorial Museum and Library, New Delhi.
21. *Ibid.*
22. Sandhu, Virendra, *Shaheed Bhagat Singh, op. cit.*, pp. 13-14.
23. *Ibid.* pp. 14-22.
24. *Transcript of an Interview with Chabil Das*, Nehru Memorial Museum and Library, New Delhi.
25. Sandhu, Virendra, *Shaheed Bhagat Singh, op. cit.*, p. 23.
26. *Transcript of an Interview with Jaidev Gupta*, Nehru Memorial Museum and Library, New Delhi.
27. *Ibid.*
28. Gupta, Manmath Nath, *Bhagat Singh and His Times, op. cit.*, pp. 97-98.
29. Deol, Gurudev Singh, *Shaheed-e-Azam Sardar Bhagat Singh, op. cit.*

2

ROOTS OF EXTREMISM

The roots of resurgent extremism may be traced from numerous factors which greatly revolutionised our thinking as well as actions. In the early phase of its development, the new youths in the form of dedicated recruits were drawn from amongst the English educated middle classes. The prevalent system of English education and consequent English educated job-oriented careers had caused much dissatisfaction, bitterness, anger and frustration amongst the youths in whose minds a feeling of inferiority complex was lurking consistently on account of uncertainty in their future life. With the spread of education, the embers of bitterness began to burn brighter among the unemployed and under employed intelligentsia.[1]

Besides, in the last few years of the nineteenth century, even nature was very unkind, cruel and destructive to the peasantry which formed a major portion of Indian population. Earlier hit by minor famines in the seventies and eighties, India had to face major famines in 1896-97 and 1899-1900 which engulfed very severely the village folk whose very subsistence depended upon the paltry produce of the soil in the normal weather and favourable circumstances. Even an able, efficient and shrewd administrator like Lord Curzon, the Viceroy of India from 1899 to 1905, failed to have a full grip over the deteriorating economic conditions caused by the severe onslaughts of famines during his time, and he estimated with much concern that one-fourth of the entire population of India had come to a greater or lesser degree within the radius of relief operation.[2]

Repelled by the victorious West and disgusted with their contemptuous attitude, the new generation, in increasing numbers, naturally sought shelter and subsistence in the womb of Mother India, in her heritage and history. The chauvinism of the West bred an equally aggressive chauvinism in the Indian mind. National paranoia, as often in history, led to national narcissism.'[3]

During this period, a new kind of feeling had developed in the minds of people which gave them clear cut inkling about various kinds of weaknesses in the body politic of the British Raj. The international situation was such that it could not boast of her friendship with a few European powers. Indeed in France and Russia, she had her traditional rivals, and Germany, under the new Kaiser, had also entered as her rival. Britain's diplomatic isolation had become a glaring phenomenon during the Boer War. The Kaiser's utterances belied the hopes of Great Britain. Above all, the defeat of Russia by Japan had greatly engendered the feeling of confidence and hope amongst Asian powers, and India was no exception in this kind of thinking that if a small nation like that of the Japanese could have an easy walk over a big power like that of Russians, it could also be possible for her to achieve this kind of success over the British Raj by devising and implementing successful revolutionary means. This kind of model had indeed emerged after a long span of about two decades after the Indian revolt of 1857-58.

In the Punjab, the revolutionary movement sprouted in the shape of Bharat Mata Society founded and organised with the initiative of Sardar Ajit Singh, Sardar Kishan Singh, Sufi Amba Prashad, Mehta Anand Kishore, Pindi Das, Kidar Nath, and last but not the least, Banke Dayal. Founded in 1907, it did function as a secret youth movement in various areas of the Punjab. Their personal contacts with the people, speeches, meetings and fatherings and patriotic literature greatly helped them in inculcating a renewed spirit of sacrifice amongst the people which prepared them for the struggle abroad.

The revolutionary activities of Babbar Akali Movement in the Punjab had significant impact on the youths of the region and the uneducated village folk which greatly helped in numerous ways for the success of this movement.[4] The rise of the movement is ascribed to the patriotic[5] activities of Sardar Mota Singh of Patara, a small village at a distance of about five kilometres from Jullundur Cantt.

It was he who called into being the party of Sikhs in the Doab who believed in violence and who hoped to make government impossible by a campaign of murder and terrorism throughout and beyond the province.[6]

Associated with Mota Singh was Sardar Kishen Singh Gargaj, a pensioned Havaldar of the 2/35th Sikh and from a village, Baring, in Jullundur district who had been involved in the Akali Conspiracy Case of 1921 in which the murders of ring of C.I.D. and of other officials were contemplated.

The only effect of this was the formation of a band of Sikh desperadoes who undertook to protect Sardar Mota Singh at the cost of their lives and who acted as his bodyguard, when he attended and spoke at meetings. No one among the public was prepared to assist in his arrest and the local police was unable to effect it for some time. Mota Singh was ultimately arrested in June 1922 and was lodged in the Nagpur Jail.[7]

In August 1922, appeared the paper called the Babar Akali Doasba. The leaflets which contained violently seditious stuff were distributed free of charge. There was no lack of sympathizers and active helpers and by the winter of 1922 an absolute reign of terror had been established in Hoshiarpur and Jullundur districts. In February, a dacoity was committed at Rahon by the Babbar Jatha and a few days later a zaildar of the Kapurthala State was brutally murdered for having assisted the State authorities in the suppression of Akali agitation. Besides, there were a few other murders and daring dacoities.[8]

The Sikhs generally sympathised with the Babbars, and the few who did not were too afraid to help the officers of the government. It was, therefore, decided to throw a wide net and simultaneous raids were made in 39 villages on 22 April 1923, and as a result of these raids 71 persons were arrested. These arrests proved a serious blow to the organization.[9] Enquiries were, however, carried to the Montgomery district where it was found that Sardar Jawala Singh and others had committed a murder for robbery and the nucleus of another Babbar organisation had been formed. This scheme was nipped in the bud by the proof of the murder and by action under the preventive section of the Criminal Procedure Code.

In January 1924, the first supplementary case with 36 accused were sent for trial, the accused included certain men of the army charged with supplying ammunition and bombs to the gang. With the death of Sardar Waryani Singh the only operation to be undertaken had been 'clearing up'. The total number of accused against whom evidence had been collected of active participation in the conspiracy was 169 of whom 8 were still at large. Of these accused no less than 54 were men who had served in the army and[8] were returned emigrants. The enquiries had resulted in the working out of 32 cases committed by the 'Babbar gangs' besides[9] miscellaneous cases of murder, dacoity, etc. Action was taken under Section 107, Criminal Procedure Code, against more than a hundred persons in the Doaba, and ninety six cases under Section 216 of the Indian Penal Code. Apart from stolen property in the shape of jewellery etc. the recoveries included four mill bombs, six shot guns, eleven revolvers, six swords, five chhavis, 285 cartridges of different kinds, and one detonator.[10]

In the second decade of the twentieth century, the scene of India's struggles and sufferings shifted to the heart of the Punjab. The region, well known for its alluvial soil, rich culture and traditions, had been the gateway of India for foreign invasions in the last few centuries. Its martial races, however, bore these onslaughts and faced them with courage, conviction and determination. The struggle against the Raj was, however, consistent and long, and it was a mixed episode tinged with non-violence and violence.

As the strains and stresses were greater in the Punjab, the agitation there was more extensive than in any other province. There was greater fervour, more strident emotions, large gatherings of crowds and all this tended to rowdy demonstrations, alarm of authorities and frequent clashes. Punjab had achieved a reputation of turbulence since the agrarian troubles of 1907. The conditions had greatly deteriorated since then. The province was feeling a sense of deep frustration...'[11]

The incident of Komagata Maru an historic event in the second decade of the twentieth century, had a romantic setting. It left the sea-shores of the Hongkong port on 4th April 1914 with one hundred and fifty five immigrants. One hundred and eleven joined them in

Shanghai, eighty-six at Moji and fourteen at Yokohama.[12] The ship reached Vancouver on 23 May. The trouble started when the nation authorities at Vancouver raised serious objection in connection with the immigration laws of the country. For example, they were asked to show their health certificates; two hundred dollars in cash and to fulfil other conditions. As the immigrants failed to do so, the Komagata Maru remained anchored and this kind of behaviour of the sea-shore officials led Indians of Vancouver to form a 'Shore Committee' which moved the Supreme Court. 'On 7 July, the Supreme Court gave its judgment that the new Orders-in-Council prevented it from interfering with the decisions of the Immigration Department, and that was the end of their hope of getting justice and legal redress of their grievances.'[13]

At this legal decision, the authorities asked the Komagata to vacate the Canadian shores. And at last it sailed from Yokohama on 23 July 1914.[14]

The functioning of the Ghadr Party in the Punjab was unique and historic in its content. Manned by most devoted, sincere and selfless workers imbued with the patriotic zeal and fervour of the highest order, the party owed its origin from San Francisco on 18 August 1913, about a year prior to the outbreak of the First World War which had deeply involved the British government and its colonies all over the world. Its Presidentship was entrusted to a sincere, intelligent and painstaking youngman, Lala Hardayal, and secretary to Sohan Singh Bhakna, a fiery speaker and devoted worker. Through their weekly *Ghadr*, they propagated their ideas for winning independence after ousting the British bureaucracy from India with a revolution. The party had much impact on thousands of emigrants from the Punjab to Canada. They were indeed unhappy over various rules and regulations which governed their very existence in a way which was repulsive in its nature.

The episode of Komagata Maru and the insults hurled on its passengers, resulting into the total incidents of about 20 Ghadrites and injuries to numerous others, added fuel to the fire. With all kinds of anger, resentment and dissatisfaction, and with no hope for the future, these Ghadrites came back to the Punjab with the blessings of their devout leader, Sardar Gurdit Singh of Sarbali village in Amritsar district. Soon their anti-Raj propaganda worked well in the rural areas and the people began to feel that a great change was coming in the near future.

The revolutionary activities of the Ghadrites included looting of banks, and government treasury; attacking police stations, railway stations and other agencies controlled and administered by the government. They established bomb factories both at Lahore and Amritsar, and committed a large number of outrages, murdered officials and wrecked trains and communications. They printed and distributed considerable revolutionary literature, and their declared object was Ghadr, i.e. revolution. In January-February 1915, they organised disturbances on a large scale in south western Punjab, and a considerable number of murders and dacoities involving much destruction of property were perpetrated. The peasantry also played its own part in this respect and some of them propagated amongst the people that the British government had come to an end and that they owed allegiance to the German Kaiser and the Amir of Afghanistan.[15]

Michael O'Dwyer in his speech in September 1917 in the Legislative Council, said as follows : 'Hon'ble Members are doubtlessly familiar with the serious dangers which menaced the security of the province during the first two years of the war, the Ghadr conspiracy and other real and covert movements, engineered by the King's enemies within or without India, with the object of subverting the Government, but, perhaps they are less familiar with the action, the prompt, vigorous and decisive action taken by the people of the Punjab, Muhammadans, Sikhs and Hindus, to range themselves on the side of law and order and to stamp out sedition and anarchy. There was no hesitation, no sitting on the fence, no mawkish sympathy with red handed crime, no insincere apology for so called misguided youths pursuing noble ideas, no subtle distinction between evolutionary and revolutionary patriotism.'

The type of violent agitational approach of the Ghadrites was taken serious note of by the government which was fully armed with a new law, known as the Defence of India Act, which armed the police and the bureaucracy with a wide range of powers to deal with the insurgency caused by them. With no strong and deep linkage with one another; lack of resources and a well organised leadership; bereft of mass support; outbreak of the war and involvement of whole nation in it, and last but not the least, the superior resources and striking power of the British government, all these factors proved suicidal for the continuance of their revolutionary activities for a long time. Consequently, during the second year of the First World War, their rendezvous were smashed very severely as a result of which many were imprisoned and convicted; some were transported for life and about twenty of them were sentenced to death.

The kind of model of self-sacrifice had deep-rooted effect on the minds of the people who were greatly worried over their economic situation which had completely broken their backbone. Their thinking and line of action, at a later stage, had a deep linkage with the repulsive prevailing situation in the Punjab.

Principal Chabil Das of National College, Lahore, and a prominent member of the Servants of the People Society, reminisced thus :

'Among Ghadrites who came back to India to arouse the people against British imperialism, Bhai Parmanand was also one of these persons, who was inspired along with Dr. Hardayal and other Ghadrites. Then he was charged with high treason and was sentenced to death, but at the intervention of Pandit Madan Mohan Malaviya and Mr. C.F. Andrews, his sentence was commuted to transportation for life. Later on, at the expiry of about five or six years, he was released.'[16]

Although crushed with an iron hand, the Babbar Akali Movement had significant impact in reawakening the peasantry and apprising them of their fundamental rights. By their speeches, writings and personal contacts with the masses of the Punjab, they succeeded in inculcating anti-Raj feelings in them and a sense of hopefulness for winning independence after ousting the British from their motherland.

The most shocking tragedy of the struggle was enacted on April 13, the Baisakhi day, the Hindu new year in the month. About 1,600 rounds were fired and firing ceased only when the ammunition was exhausted. Hundreds more were injured and left to die without medical attendance or even water. This indeed was the most shocking tragedy of the freedom struggle in the twentieth century.

'Martial Law was imposed on the whole of the Punjab. What followed was a reign of terror. All leaders of the people, whether in political life or not, were arrested and thrown into prison. Every effort was made to humiliate the people. In one street in Amritsar, they were compelled at the point of the bayonet to crawl on their bellies. There was strict censorship of the press. No reports of the happenings were allowed to go out of the province. But as days passed, rumours about the diabolical doings of the authorities spread throughout the country.'[17]

'The Amritsar massacre caused an unparalleled indignation not only in India but also among all fair minded people in England. But the steps taken by the Government to compensate the sufferers and to punish the wrong doers were extremely inadequate.'[18]

E.S. Montagu was fully aware of the fact that the policy of ruling India by the sword alone was impossible and he had the realisation that the government, foreign or indigenous, could not rule a people for any length of time without the willing support of unquestioning acquiescence of a section of the effective classes of the people. Besides, he had a personal stake in the Indian affairs. He had worn himself out to process the reforms and was naturally most anxious for their success. He made up his mind and in the budget debate on May 22, gave a promise to hold an enquiry.'[19]

The reaction of the people rose to phenomenal height. The *satyagraha* pledge to be taken by the people said, "Being consciously of opinion that the (Rowlatt) Bills ... are unjust, subversive of the principles of liberty and justice, and destruction of the elementary rights of individuals on which the safety of the community as a whole and the State itself is based, we solemnly affirm that, in the event of these Bills becoming law and until they are, withdrawn, we shall refuse civility to obey these laws and such other laws as a committee to be hereafter appointed may think fit, and we further affirm that, in this struggle, we will faithfully follow truth and refrain from violence to life, person or property."

Although Punjab was isolated and cut off from the rest of India demands were made for the recall of O'Dwyer and

Chelmsford. Indemnity for the prisoners was pressed and enquiry in the Punjab happenings urged, both in England as well as in India. During this period the leaders of the Home Rule League and the Liberal Federation showed activity in giving evidence before the Joint Parliamentary Committee. Among them were Vithalbhai Patel, Bal Gangadhar Tilak, Bipin Chandra Pal, Mrs Annie Besant, Surendranath Banerjee, Bahadur Sapru, Srinivas Sastri and others. They all impressed upon the Secretary of State the urgency of an enquiry into the recent events in order to pacify public opinion.

The Bharat Naujawan Sabha, the Communist Party, the Kirti Kisan Party, the Socialist Party, the Students Unions and the Bal Bharat Sabha played their role in the freedom struggle. Undoubtedly Youth Association of Bengal did a commendable work in this regard. After his return from England in 1921, Subhas Chandra Bose to whom the present movement owes its origin to a great extent put in his efforts to bring together all young students in schools and colleges on a common platform and inculcated in them the nationalistic and patriotic fervour for the country. Later on, when he gave a call to young girls to join it, its name was changed to "All Bengal Youth Association." Its main aim was to consolidate and extend its power both by open propaganda and secret organisation. It was also to work for the eradication of social evils and uplift of the depressed classes. Bose personally appealed to youngmen and was able to secure their support in all 'open and secret' work. He was of the opinion that 'Youth movements are not reformatory but revolutionary. The youths of India are no longer content with handing over all responsibility to their older leaders and sitting down with folded hands or following like dumb driven cattle. They have realised that it is for them to create new, free, great and powerful India.'

Besides, the All-Bengal Students' Association, formed in 1924, had a well disciplined organisation and cadre of young workers in various areas of the province. Pramod Kumar Ghosalan and Birendra Kumar Das Gupta were elected its President and General Secretary. The Association aimed at inculcating the spirit of selfless, sacrifices and patriotism amongst the youths of the province and also to spread its objects beyond the borders of Bengal.

The work of the revolutionaries was wholly secret. The prime function of the leadership was to spot out potential, sincere and devoted comrades and then pursue them relentlessly with literature and arguments till they gave their assent to join them. They took precaution not to approach youths with a bad reputation or a family background which had no reputation in accordance with their standards.[20]

"So revolutionaries had to be really handpicked. Ugly surprises would always occur, nevertheless we had to take certain precautions. When I spotted out a possible comrade, I gave him books. In the beginning historical novels with a patriotic tilt, such as Bankim's and Ramesh Chandra Dutt's novels and patriotic poems of Navin Sen, Maithilisharan and Rajni Sen were given .... I gave him lives of patriots like those of Rana Pratap, Shivaji, Garibaldi, Mazzini, Wallace, etc... The life of the Indian revolutionary martyrs constituted the last course. In between the youth was fed on the history of the Chinese Revolution (1912), Zaghlul Pasha's struggle for Egyptian independence, etc."[21]

'Prolonged discussions took place in our ranks about what to do to break this stagnant calm. Socialist literature was trickling in the triumph of the November revolution, the consolidation of the socialist regime in Russia and more than anything else, the aid given by the Soviet Union to Asian countries like Turkey and China against imperialist powers attracted us towards the new socialist state and towards the ideas and principles it embodied.'[22]

Bhagat Singh was indeed very active in the Punjab. With his efforts and cooperation of his comrades was formed the *Naujawan Bharat Sabha*, a militant youth organization to propagate socialist ideas and above all preach anti-Raj statements. The Sabha became a very popular political organ which played a significant role in the radicalisation of the youths of the country.

The background of the events which had led to the martyrdom of Bhagat Singh and raised him to the pantheon of India's greatest heroes was provided by the general despondency among the youths in India after the suspension of the Non-cooperation Movement in February 1922. The nationalist movement seemed to be paralysed on the one hand by dissensions among the top leaders, and on the other, by the growing communal bickerings and riots.

It was in this atmosphere that many youngmen in various parts of the country came to the conclusion that the methods followed by Gandhi and the Congress were not capable of carrying the country forward and preparations must be made for an armed struggle to wrest power from British hands. And in order to prepare the people for such a struggle it would be necessary to carry on propaganda through brave deeds of heroism and effective protest against the barbarities perpetrated by the British.[23]

Bhagat Singh's ancestors were Sandhu Jats and were settled since generations in a small but prosperous village, Khatkar Kalan, situated at a distance of about thirty kilometers from Phagwara railway station. Geographically, it is in the Jullundur district. The Doab was a renowned belt of revolutionaries and was always humming with patriotic fervour. Khatkar Kalawas, now a place of pilgrimage, could not remain in isolation and with the passage of time, the spirit of selfless work, patriotic fervour and supreme sacrifice for the country was kindled in a small house of this village.[24]

In the first half of the nineteenth century, some members of Bhagat Singh's family had served in various capacities in the army of Maharaja Ranjit Singh and had shown distinction in their service careers. They were well conversant with the onslaughts of Pathans from North West Frontier as well as the speedy expansion of the British commercial and political interests in India. And above all, the British annexation of the Punjab and the subsequent administration in the province had deep imprint on the patriotic fervour of his ancestors.

Like any other village of his time, Bhagat Singh's grandfather, Sardar Arjan Singh led a simple life, his main avocation being the medical practice of Unani system. His clients were the common village folk who had complete faith in the efficacy of his medicines and his verbal advice to be followed with meticulous discipline by the patients. Imbued with the knowledge of Urdu, Punjabi, Persian and Sanskrit, the doctor kept himself fairly busy in various kinds of social and religious work in his area, which made him a popular figure.

During this time, the Arya Samaj movement, with its firm socio-religious base, had gained much momentum in many cities and towns of North India. To be its member or worker was in fact a big gain in the social life of any person. So Arjan Singh became a member of this organisation and showed so much activity in it that he preached its gospel amongst the simple illiterate villagers. His powerful influence coupled with the reforming zeal had much impact in the suburbs.

In the first decade of the 20th century, the Punjab had a broad spectrum of popular leaders working in the state. Two such leaders were Sardar Ajit Singh and Sardar Kishan Singh, Bhagat Singh's paternal uncle and father. Interestingly, although middle class, both leaders were in great opposition to the mainstream leadership of the Indian National Congress, particularly Lala Lajpat Rai. Both brothers were consistently radical in attempting to mobilise the masses to oppose the British at every opportunity that arose.

In an article called 'Emergence of the Punjab in the Freedom Movement',[25] Bhagat Singh traces their inspiration to the extremist leader, Bal Gangadhar Tilak. 'Having seen their enthusiasm at the 1906 Congress Convention in Calcutta, Lokmanya was pleased and in bidding them adieu, gave them the responsibility of strengthening the movement in the Punjab.' On returning to Lahore, both brothers 'started a monthly newspaper called *Bharat Mata* to propagate their ideas'. Since they had no money and no influence among the rich, says Bhagat Singh, they had to collect everything necessary for propaganda work themselves. This they did by attracting a crowd in the market by ringing a bell and giving a lecture 'on how foreigners had destroyed India's industry and commerce'. This was followed by an announcement that an important meeting would be held within the week at the *Bharat Mata* office. After the first two meetings held on two consecutive Sundays, a decision was taken to hold a meeting every Sunday. Many people joined the group including Lala Lalchand 'Phalak', the 'national poet' of the Punjab, Lala Pindidas, Dr. Ishwari Prasad and Sufi Amba Prasad. Sardar Ajit Singh and Mahant Nandkishore were elected the President and Secretary of the group now organised as the 'Bharat Mata Society'.

In 1887, the Punjab government created the Chenab Colony, by diverting the Chenab river into a system of perennial canals, rapidly turning the barren wasteland of the Central Punjab into fertile farmland.[26] The colony was to be a model for the rest of Punjab, supervised by a paternalistic administration. 'Healthy agricultural communities of the "best Punjabi type" would be established... (which would) in turn demonstrate to other Punjabis how proper sanitation, careful economic planning and co-operation with the government could result in a higher standard of living'.[27]

In October 1906 the government introduced the Punjab Colonisation of Land Bill in the local Legislative Council. This Bill established retroactive conditions concerning sanitation, tree planting and construction in the Chenab Colony. In case of breach of the conditions, fines would be collected in the same fashion as revenue. Section 31 ruled that courts would have no jurisdiction in the Colony. In November, the government announced a drastic increase in the occupier rate (charge on canal water). In the Bari Doab canal area running through the districts of Amritsar, Gurdaspur and Lahore, rates had been lower because the government had hoped that a leniency in revenue would ensure the loyalty of the Sikh Jats who supplied recruits for the Indian Army.[28] The yeomen farmers of the Colony were already discontented over the maladministration in the Colony and opposed the extra legal fees that the Bill sought to legitimise. Graft and corruption had galvanised the whole Colony into opposition to the entire system of interference and paternalism. From 1903, Sifaj-ud Din Ahmed, a retired postal officer, started the *Zamindar*, a newspaper, to publicise the colonists' plight. It was only a matter of time for discontent to turn into political opposition.

Bhagat Singh's grandfather, Sardar Arjan Singh, had migrated to the canal area and settled at Banga, Lyallpur. It is, therefore, not surprising that Sardar Ajit Singh was able to articulate so clearly the problems of the colonists and the objection to the Bill. On 22nd and 23rd March, 1907, the *Zamindar* held a public meeting to protest against the bill. Although this was a platform for the rich yeomanry, Ajit Singh sent delegates from the 'Bharat Mata Society' to launch an agitation against the British. Lala Lajpat Rai was asked by the newspaper to give a speech. Bhagat Singh in his article on the national movement in the Punjab shows how Lajpat Rai and Ajit Singh differed on this issue:

"Before leaving (for Lyallpur) Lalaji sent a message to Sardar Ajit Singh saying that the government should be thanked for (a previous) amendment and then asked to repeal the law." To this Ajit Singh is said to have replied, "We shall prepare the masses for a no revenue campaign. Also we can never thank the government."

The meeting itself is described by Bhagat Singh thus, "Lalaji was received by a large rally and consequently reached the *pandal* two hours late... In the meanwhile, Sardar Ajit Singh gave a speech. He was an impressive speaker. His tireless style of speech made the audience enthusiastic and by the end he had a large following of people. By the time Lalaji reached the *pandal*, the masses were with the Bharat Mata Society... Lalaji was Punjab's finest orator but the style, the fearlessness and determination with which he spoke was something else. He received an ovation after every line. After the meeting many people dedicated their lives to the motherland".[29] This account is also corroborated by N. Gerald Barrier, "Lajpat Rai attempted to be moderate, but as happened frequently with his speech making, the crowds' frenzy drove him to use phrases and ideas verging on what the British termed 'sedition'. After the meeting, Lajpat Rai went on a lecture tour in the United Provinces, while Ajit Singh began to organise the farmers of Amritsar and Lahore. Under Ajit Singh's leadership the colonists passed a resolution supporting a boycott of British goods and started a campaign to ensure that nobody paid the new water rate. The price of disobedience was social ostracism by the offender's caste or a fine of Rs. 500."[30] The Lieutenant Governor of the Punjab, Denzil Ibbetson believed sedition in the province to be taking two directions. First Ajit Singh was trying to spread disaffection among the troops and the students, and secondly the 'fomenters of unrest' were corrupting the yeomanry. He asked the Government of India for permission to deport Ajit Singh and Lajpat Rai so as to be able to 'strike terror into the minds of those concerned'. Ajit Singh was deported to Mandalay Jail. Sardar Kishan Singh, though less prominent than his brother, was also part of this agitation and was put into jail.

Educated in Sain Das Anglo Sanskrit High School at Jullundur, indeed one of the premier Arya Samaj institutions in the Doab, Sardar Kishan Singh had imbibed, a deep sense of nationalism at a very young age from the renowned teacher and a selfless Arya Samaj worker, Lala Sundar Dass. This type of education with its main emphasis on simplicity, dedication, sacrifice and love for his country paid him rich dividends and he, later on ingrained such feelings in each and every member of his family.

Soon after passing out of school, Kishan Singh engaged himself in a private job which fetched him a bare pay enabling him to maintain his family with honest means worthy of the tradition of his ancestors. The well established Industrial Prudential Company with its headquarters in Bombay offered him an opportunity of service in its branch located in Punjab. But this kind of work merely for his subsistence and maintenance of the members of his family did not interest him much. Instead, he involved himsalf in more fruitful activities, both social and political in the years to come.

Thûs, Kishan Singh proved himself instrumental in helping and organising relief work for the backward famine-stricken people of Central Provinces and the victims of the severe earthquake in Kangra. He also worked shoulder to shoulder in helping Lala Lajpat Rai for the cause of the destitute and helpless orphans. His involvement in social sphere soon crossed the slow moving borders into an active field of political activities for which the family was destined in the coming few decades of the twentieth century. So in 1906, all the three brothers had a meeting with Sufi Prasad, Sardar Kartar Singh`and Mahasha Ghaseeta Ram and decided to raise their patriotic fervour against the condemned and much criticized Colonization Act and the Bari Doab Canal Act. These acts of the government had greatly damaged the interest of the peasantry in the canal areas of the Punjab and it did need the services and sacrifices of more and more prominent, sincere and dedicated persons for its cause in the movement already launched with full force. The foundation of political party known as Bharat Mata Society with the help and cooperation of Sufi Amba Prasad and others offered him a platform for his political and agitational approach against the measures of the government. Next year, when he took an active part in the anti-government agitation in connection with the peasantry of the canal region, he found himself in jail along with numerous colleagues of the movement.

Kishan Singh's political activities were intensified with the passage of time. During the First World War the activities of Ghadrites were stepped up on the Indian political horizon. As expected of him, he did not hesitate to support this anti-government movement in every possible way. So far so that Kartar Singh Sarabha, one of the prominent members and hero of the Ghadr movement, reported about the receipt of one thousand rupees in the form of financial support, coupled with his other political activities which were thought to be anti-government, which could not remain a secret for a long time. The police unearthed this kind of activity sponsored and supported by Kishan Singh at a time when the Ghadr movement was being crushed by the Punjab government. By this step of the government, Kishan Singh also became the victim of police rules and regulations. He was, therefore, labelled as a dangerous element and a supporter of anti-government agitation and movement. For all this, he was punished; he was interned in his village for some time during the First World War. Thus his political activities which had patriotic fervour brought him numerous personal troubles both in jail and outside it. His sacrifice for the motherland and above all the family tradition maintained by his younger brother, Sardar Ajit Singh and his son Sardar Bhagat Singh are worthy of emulation.

Sardar Kishan Singh used to live in Nawan Kot, Lahore, and from there he shifted to Kucha Nihal Chand in 1920. The atmosphere in his house, according to Jaidev Gupta, a class fellow of Bhagat Singh, became very tense after the tragedy in Jallianwala Bagh as all types of people used to come there and talked about martial law and its effects on the people of the Punjab.[31] These prominent leaders were Bhai Parmanand, Kedar Nath Sehgal, Lala Pindi Das and others.[32]

It is a strange coincidence that when Bhagat Singh was born, his father who was lodged in Lahore Central Jail in connection with an agitation against the Colonization Bill and his uncle, Ajit Singh, whose sacrifice for the national cause is proverbial, was in Mandalay Jail, both were released on the same day. It is commonly believed that the child Bhagat Singh was initially named as Bhaganwala by his grand mother as the birth of the new baby brought good luck and fortune to the members of the family. Later on, he was named Bhagat Singh. From the very childhood, he had the signs and traits of a great man who later became an unforgettable hero of the Indian nation.

REFERENCES

1. Bose, A.C., *Indian Revolutionaries Abroad, 1905-1922* (Patna: B.R. Publishers), 1971, pp. 12.
2. *Lord Curzon's Speeches, 1898-1905*, Vol. II, pp. 94-105.
3. Bose, A.C., *Indian Revolutionaries Abroad, 1905-1922, op. cit.*, p. 5.
4. Home Poll File 200, 1926.
5. It is a typical bureaucratic phraseology.
6. Home Poll File 200, 1926.
7. *Ibid.*
8. *Ibid.*
9. *Ibid.*
10. *Ibid.*
11. Chand, Tara, *History of the Freedom Movement in India*, Vol. III (New Delhi: Publications Division, Ministry of Information and Broadcasting, Government of India), 1972, p. 478.
12. Of them only 25 were Muslims and rest were Sikhs.
13. Bose, A.C., *Indian Revolutionaries Abroad, 1905-1922, op. cit.*, p. 63.
14. Khushwant Singh and Satindra Singh, *Ghadr, 1915* (New Delhi: Penguin),1966, pp. 28-29.
15. About 28 persons were hanged and 100 transported.
16. Transcript of Interview with Chabil Das, Nehru Memorial Museum and Library, New Delhi.
17. Kripalani, J.B., *Gandhi : His Life and Thought* (New Delhi: Publications Division, Ministry of Information and Broadcasting, Government of India), 1970, pp. 82-83.
18. Bose, Subhas Chandra, *The Indian Struggle, 1920-1942* (Bombay: Allied Publishers), 1967, p. 26.
19. Chand, Tara, *History of the Freedom Movement in India*, Vol. III, *op. cit.*, pp. 481-482.
20. Gupta, Manmathnath, *They Lived Dangerously : Reminiscences of a Revolutionary* (Delhi: Peoples Publishing House), 1969, p. 92.
21. *Ibid.*, p. 93.
22. Ghosh Ajoy, "Bhagat Singh and His Comrades" in *New Age*, 22 March 1931.

23. Prasad Bimal, "Remembering Bhagat Singh" in *Mainstream*, 28 March 1981.
24. Sandhu, Virendra, *Amar Shaheed Bhagat Singh, op. cit.*, p. 1.
25. Bhagat Singh 'Emergence of the Punjab in the Freedom Movement', written in Urdu. Published in the 1931 weekly *Bandematram*. Reprinted in Jagmohan Singh and Chaman Lal (eds.) *Bhagat Singh aur Unke Sathion Ke Dastavez*, Delhi, 1987.
26. Barrier, N. Gerald, 'The Punjab Disturbances of 1907: The Response of the British Government in India to Agrarian Unrest', in David Hardiman (ed.), *Peasant Resistance and the Raj*, New Delhi, 1992.
27. *Ibid.*
28. *Ibid.* Approximately 30,000 Sikhs (23% of the Army) and 18,000 Punjab Muslims (13% of the Army) were in the Army, p. 245.
29. Bhagat Singh, "Emergence of the Punjab in the Freedom Movement", written in Urdu. Published in the 1931 weekly *Bandematram*. Reprinted in Jagmohan Singh and Chaman Lal (eds.) *Bhagat Singh aur Unke Sathion Ke Dastavez, op. cit.*
30. Barrier, N. Gerald, "The Punjab Disturbances of 1907: The Response of the British Government in India to Agrarian Unrest", in David Hardiman (ed.), *Peasant Resistance and the Raj, op. cit.*
31. Transcript of Interview with Jai Dev Gupta, Nehru Memorial Museum, New Delhi.
32. *Ibid.*

3

POLITICAL IDEOLOGY

In his early childhood Bhagat Singh was greatly influenced by the political activities of his grandfather, father and two uncles. Among them his uncle Ajit Singh impressed him the most. The events related to Kamagata Maru, Ghadr rebellion and the martyrdom of Kartar Singh Sarabha also caught his imagination. In 1914, the enforcement of the Martial Law in Punjab and the Jallianwala Bagh tragedy (1919) made him a determined opponent of the British rule in India. During his studies in school and later on at the National College, Lahore, he read several books on French Revolution, Marxism, Russian Revolution and a few other revolutions of the world. At that time he began to contemplate seriously on the path of revolution.[1] Greatly influenced by the anarchism theory of Bakunin, Bhagat Singh was attracted towards anarchism, but under the influence of Chabil Das and Sohan Singh Joshi he gradually got himself rid of that theory and became a revolutionary socialist.[2]

At Kanpur he came into contact with several revolutionaries of northern India. Along with it he studied the history of revolutions which occurred in different parts of the world. He also read communist literature. By the time he returned to Lahore, Bhagat Singh had decided his future course of action. He became an active member of the Hindustan Socialist Republican Association. Even after becoming an active member of the revolutionary party and the *Naujavan Bharat Sabha* of Lahore, Bhagat Singh spent considerable time in studying revolutionary literature on Marxist, socialism, anarchism and revolutionary socialism. Besides, he read the works of Plato, Aristotle, Socrates, Hobbes, Mill, Engels, Mark

Twain, Rousseau, Rabindra Nath Tagore, Ghalib and a few other authors. During his confinement in the jail he continued his readings. Several books were brought for him by his brother and friends. In the diary kept by him in the jail he scribed some of the extracts from the books read by him. The themes covered by him were intensely diversified in their nature.[3] The themes noted in the above diary relate to freedom from poverty, origin of the institution of family, savagery, development of civilization, marriage as an institution and its deficiencies; power of coercion, origin of the state religion; benevolent despotism, British rule in India; liberty and its concept among English people; monarch and concept of monarchy, natural and civil rights, rights of labour; anarchism, condition of labour class; morality; hunger; freedom; alms; liberty; free thought; oppression; martyrdom; charity; prison and the prisoner; death; child labour; capitalism; social institutions; revolution and classes; democracy; reforms; aims of socialism; aims of world revolutions; man and machinery; imperialism; dictatorship; bourgeois democracy; concept of leadership; arbitrariness; political party; aims of communists; tactics and strategy; Marxism *versus* Socialism; law of justice; kind of law; political crimes and punishments; foreign subjection; life and education; theory of Divine Right of Kings, French Revolution; Constitution of free India; untouchables; India and China, rural and urban problems and Arya Samaj and so on. Besides, various kinds of statistics were noted in it.

But instead of being confused by such a vast reading, Bhagat Singh was clear in his mind about the subjects studied by him. His clarity of thought about the various subjects read by him is clearly discernable from his writings in the jail. It shows that he was more or less a scientific revolutionary who had the capacity to mould his thoughts through reading and contemplation. As, for example, he admits that in the beginning he believed in terrorism but as time went on, he realised that killing of persons was senseless if it targeted the innocent ones. It only aggravated violence and side tracked the issue of social transformation. But force was to be met with force. Force when aggressively used was violent and was morally unjustifiable but when it was used in the furtherance of a legitimate cause, it had its moral justification.[4]

Bhagat Singh firmly believed in Marxism. He believed in the maxim advocated by Marx that political actions were not the cause but were the products of economic forces. A change in the balance of economic power was the rationale upon which depended all other changes of human history. According to Bhagat Singh, the struggle for independence in India was basically a struggle for improving the economic status of the exploited masses. If in independent India poverty was not removed the country would only be free in name. If independence meant only a change of masters it was against the interest of the common man.

Bhagat Singh was a votary of revolution. He had coined the slogan 'Long Live Revolution' (Inquilab Zindabad). In the manifesto of the Hindustan Socialist Republican Association and a pamphlet issued by it in 1925,[5] which were written by Bhagat Singh and the various statements issued by him during his trials in the courts, he tried to explain the concept of revolution as was visualised by him.

In the manifesto referred to above, he outlined a guideline for those who volunteered themselves to work for bringing about a revolution. It says, "upheavals have always been a terror to holders of power and privilege (but) Revolution is a phenomenon which nature loves and without which there can be no progress either in nature or in human affairs. Revolution is not a philosophy of despair or a creed of desperadoes. Revolution may be anti-government; but is certainly not anti-man. It is a vital living force which is indicative of eternal conflict between the old and the new. There is no concord, no sympathy, no rythm within Revolution. The music of the 'spheres' of which poets have sung would remain an unreality if a ceaseless Revolution were to be eliminated from the space. Revolution is Law, Revolution is Order and Revolution is Truth."

In fact the revolutionaries were not bothered about disorder chaos, or anarchy which generally occurred as a glaring phenomenon during a revolutionary upsurge. According to them, such conditions were necessary for the birth of a new star. The birth of life is always accompanied by agony and pain.

While throwing bombs in the Central Legislative Assembly both Bhagat Singh and B.K. Dutt shouted the slogan, "Inquilab Zindabad". Again during the trial of Delhi Assembly Bomb Case they raised this slogan. On being asked by the judge as to what they meant by the word 'Revolution', Bhagat Singh on his behalf and that of B.K. Dutt, submitted a statement in the court. In this statement he explained the ideology of Revolution in the following words :

"Revolution does not necessary involve sanguinary strife nor is there any place in it for individual vendetta. It is not the cult of the bomb and the pistol. By 'Revolution' we mean that present order of things, which is based on manifest injustice, must change. Producers or labourers in spite of being the most necessary element of society, are robbed by their exploiters of the fruits of their labour and deprived of their elementary rights. The peasant, who grows corn for all, starves with his family, the weaver who supplies the world market with textile fabrics, does not have enough to cover his own and children's bodies, masons, smiths and carpenters who raise magnificent palaces live like pariahs in the slums. The capitalist and exploiters, the parasites of society, squander millions on their whims. These terrible inequalities and forced disparity of chances are bound to lead to chaos. The state of affairs cannot last long, and it is obvious that the present order of society in merry making is on the brink of a volcano."

"The whole edifice of this civilization, if not saved in time, shall crumble. A radical change, therefore, is necessary and it is the duty of those who realize it to reorganize society on the socialistic basis. Unless this thing is done and the exploitation of man by man and of nations by nations is brought to an end, sufferings and carnage with which humanity is threatened today cannot be prevented. All talk of ending wars and ushering in an era of universal peace is undisguised hypocracy."

"By 'Revolution' we mean the ultimate establishment of an order of society which may not be threatened by such breakdown, and in which the sovereignty of the proleteriat should be recognised and a world federation should redeem humanity from the bondage of capitalism and misery of imperial wars."

"If, however, it goes unheeded and the present system of Government continues to be impediment in the way of natural forces that are swelling up, a grim struggle will ensure involving the overthrow of all obstacles, and the establishment of the dictatorship of the proleteriat to pave the way for the consummation of the ideal of revolution. Revolution is an alienable right of mankind. Freedom is an imperishable birth right of all. Labour is the real sustainer of society. The sovereignty of the people is the ultimate destiny of the workers."

"For these ideals and for this faith, we shall welcome any suffering to which we may be condemned. At the altar of this revolution we have brought our youth as an incense, for no sacrifice is too great for so magnificent a cause. We are content, we await the advent of Revolution 'Long Live Revolution.'"[6]

Ramanand Chatterjee, the editor of *Modern Review*, Calcutta ridiculed the slogan of 'Inquilab Zindabad'. He asked Bhagat Singh its exact meaning. In December, 1929 in an editorial of the journal, he wrote, "... when a desire is expressed for revolution to live long, it is desired that the revolutionary process should be at work every hour, day, week, month and year of our lives. In other words, are we to have a revolution as often as possible ?" He further wrote, "A ceaseless revolutionary process would make India like what James Russell Lowell called the 'Catherine Wheel Republics of South America' of his days. In spite of changes after a revolution it does not produce a final state of improvement. It has to be brought about by evolution. Therefore, the slogan of 'Inquilab Zindabad' was a symbol of bloodshed and chaos."

In response to the above editorial, Bhagat Singh on his and Dutt's behalf wrote a letter to Chattelee :

"We are not the originators of this cry. The same cry has been used in the Russian Revolutionary Movement. The famous socialist writer Upton Sinclair in his novels 'Boston' and 'Oil' has used these slogans through the mouths of some of his anarchist revolutionary characters. It does not at all mean that the sanguinary strife should continue or that nothing should ever be stationary even for a short while. By long usage this cry had achieved a significance which may not be quite justifiable from the grammatical or the etymological point of view, but nevertheless we cannot separate it from the association of ideas connected with it. All such slogans are indicative of a recognised meaning which has cropped up to a certain extent..."

In Bhagat Singh's opinion the raising of the slogan of 'Jatindra Nath Zindabad' was meant to inspire the people to follow his ideals for which he suffered untold miseries and offered a sacrifice of such a great magnitude. The raising of this slogan also indicates a desire of the revolutionaries to follow the revolutionary idea with zeal.

Therefore, one should not try to understand the significance of the slogan of 'Inquilab Zindabad' only by its literal meaning. In the eyes of revolutionaries it is now a famous slogan.

He then repeated his view expressed in the statement submitted before the trial court in Delhi about revolution. Revolution does not invariably mean an armed rebellion. Sometimes bombs and pistols may become instruments for the success of revolution but only for this reason bombs and pistols cannot be synonymous with a revolution. A revolt cannot be called as a revolution although it is possible that the ultimate result of a revolt may be a revolution.

Bhagat Singh further argued, "The sense in which the word revolution has been used in that phrase is the spirit, the longing for a change for the better. People generally get accustomed to the established order of things and begin to tremble at the very idea of a change. It is the lethargic spirit that needs to be changed by revolutionary spirit. Otherwise degeneration gains upper hand and the whole humanity is led astray by the reactionary forces. Such a state of affairs leads to stagnation and paralyses human progress. The spirit of revolution should always permeate the souls of humanity so that revolutionary forces may not accumulate (strengthen) to check its external march. Old order should change, always and ever, yielding place to new so that one 'good' order may not corrupt the world. It is in this sense that we raise the slogan 'Long Live Revolution".[7]

During his confinement in Lahore Central Jail, Bhagat Singh sought an opportunity again to explain some of his ideas on revolution. In the introduction of 'Dreamland', a poem written by Ram Saran Das, he expressed the view that all the political movements of India which had hitherto played an important role in modem history were not clear about their ideals. Except for the 'Ghadr' party no other party had clear ideas as to what they were fighting for. In his opinion the 'Ghadr' party clearly stated that it wanted to replace the existing government by a republican government on the pattern of the United States of America. All other parties have only one idea i.e. to fight against the alien rulers. This idea is quite laudable but cannot be regarded as revolutionary. Revolution does not merely mean an upheaval or a sanguinary strife. It necessarily implies the programme of systematic reconstruction of society on a new and better adopted basis after delicate destruction of the exciting state of affairs, i.e. regime.[8]

According to him the liberals in India wanted some reforms under the present government while the extremists demanded a bit more and were prepared to adopt radical means for achieving their objects. The revolutionaries have always advocated the overthrow of the foreign domination. In order to drive out the foreigners from their country violent means were necessary. In this context he quoted Lenin, who said to Gorky, a renowned Russian author, "the revolutionaries have to resort to violent means as a terrible necessity."[9]

In his introduction to 'Dreamland' Bhagat Singh did not agree with the views of Ram Saran Das about the salient aspects of society to be built up after a revolution. Ram Saran Das had written that the communist society which the revolutionaries wanted to built should have charitable institutions for helping the needy and poor. But in Bhagat Singh's opinion there would be no needy and poor in such a society. He also did not agree to the suggestion of Ram Saran Das that after revolution the public servants should work in factories and farms for four hours daily. In the theory of Bolshevism, Bhagat Singh wrote, mental work was also regarded as productive as the manual labour. In a socialistic society the relations between the different elements of the society are to be adjusted on the basis of equality. The producers and distributors should be considered as equally important. The only difference which a socialist society desires is that the manual workers should not be regarded as inferior to the mental workers.

In his poem Ram Saran Das had stressed the need of the process of evolution. According to him this could be achieved by educating the masses on definite lines. But Bhagat Singh did not consider it possible to educate the masses. In the existing circumstances the British Government was not going to help any such effort. To the contrary it will suppress it mercilessly. Hence the revolutionaries were striving to capture power in order to employ all its resources for educating the masses. This process was adopted in Soviet Russia. Only after the capturing of power peaceful methods could be used to overcome the obstacles.[10]

As a student in Lahore, Bhagat Singh was greatly impressed by the mass movement launched by the Congress under the leadership of Gandhiji. He even acted as a volunteer in the Non-cooperation Movement. But he was completely disillusioned with Gandhiji when he postponed the movement soon after the incident of Chauri Chaura. According to Bhagat Singh, by the postponement of the Non-cooperation Movement India had lost a great opportunity to bring the enemy to its knees. Some time later he told his father that he could neither understand Gandhi's political strategy nor his mental approach which had hacked the movement to pieces with a single blow. Non-violence was all right up to a certain point. It had helped in building a popular response. But he was unable to understand or appreciate the strategy or morality of an act which dealt a fatal blow to a popular movement at its peak. In his opinion, an alien rule can be defeated not by gloves, but by an iron fist. At times it was inevitable. Force when aggressively applied was violence and morally unjustifible but when it was used for a legitimate cause, it had moral justification. He did not agree with Gandhiji that right results would not come from wrong methods. It was the goal which mattered. If the people gained their freedom by force, these actions were justified.

He considered non-violence as an excuse for inaction, a cover for cowardice. He, therefore, did not believe in the creed of non-violence. The kind hearted philanthrophy of Gandhiji was not required. The British must fear India.

Bhagat Singh and his comrades regarded Gandhi as a visionary. In their opinion he was not a person who could scare the British bureaucrats and politicians. Yet they had to reckon with his leadership because teeming millions followed him.

In 1928 Bhagat Singh and his associates did not support the demand of the Congress for the grant of Dominion Status to India. They wanted full independence for the country. Still they decided to respond to Gandhiji's call to boycott Simon Commission. As revolutionaries they believed that any move to stir people was a step in the right direction. Such a step, however small, would make the nation conscious of its shackles it wore. Action was the soul success in revolution.

Bhagat Singh had no faith in the policies of the Congress. According to him the Congress was still in the hands of the middle class. It wanted to secure rights for the rich by putting pressure on the government. As far as the millions of workers and farmers were concerned, they did not figure at all in the reckoning of the Congress. He wanted the farmers, labourers and the common man to be in the forefront of the struggle for independence. According to him the Congress leaders were not broadening the mass base of the party's membership because they were afraid of revolution. This type of responsibility has, therefore, fallen on the shoulders of revolutionaries.

In 1929 during the trial of Bhagat Singh and B.K. Dutt in the Delhi Assembly Bomb Case Gandhiji while appreciating the bravery of both the revolutionaries denounced violence as a weapon in the struggle for freedom. In some of the articles written by him in *Young India* he criticised the revolutionaries for their cult of violence. In an article in *Young India* entitled 'Cult of the Bomb' Gandhiji emphatically denounced the use of the bombs and pistols. Along with Gandhiji, several leaders of the Congress criticised the violent methods of the revolutionaries. Several newspapers also joined them on the issue. In order to dispel their doubts about the ideals and methods of the revolutionaries in India an article under the caption of 'Philosophy of the Bomb' under the pseudo name of Kartar Singh was circulated by the revolutionaries.[12] According to Jatindra Nath Sanyal, a comrade of Bhagat Singh, the article written by Bhagat Singh was smuggled out of the jail and was distributed as a pamphlet throughout the country on January 26, 1930.

In this pamphlet the attitude of the Congress towards the revolutionaries was discussed. Along with it the ideals and methods of the revolutionaries were also highlighted. According to the pamphlet "When a revolutionary believes certain things to be his right, he asks for them, pleads for them, argues for them, wills to attain them with all the soul force at his command, stands the greatest amount of suffering for them, is always prepared to make the highest sacrifice for their attainment and also backs his efforts with all the physical force he is capable of." He further wrote that for the acceptance of truth, soul force alone is not sufficient; sometimes physical force has to be added to it. Hence the revolutionaries stand for winning independence by all forces, physical as well as moral, at their command.

The revolutionaries firmly believed that the deliverance of the country would come only through revolution. The revolution, they are constantly working and hoping for will not only express itself in the form of an armed conflict between the foreign government and its supporters and the people, it will also usher in a new social order. The revolution will ring the death knell of capitalism, class distinctions and privileges. It will bring joy and prosperity to the starving millions who are seething today under the terrible yoke of foreign and Indian exploitation. It will bring the nation into its own. It will give birth to a new state and a new social order. Above all, it will establish the dictatorship of the proletariat and will for ever banish social parasites from the seat of political power.

In the pamphlet, the advent of revolution in the restlessness of youth, in its desire to break free from the mental bondage and religious superstitions that were holding them, was visualised. As the youth will get more and more saturated with the psychology of revolution, it will have a clearer realisation of national bondage and an intense and unquenchable thirst for freedom. This feeling of bondage, this insatiable desire for freedom will grow till, in their righteous anger, the infuriated youth will begin to kill the oppressors. Thus has terrorism been born in the country. It is a phase, a necessary and inevitable phase of the revolution. Terrorism is not complete revolution and revolution is not complete without terrorism. This thesis can be supported by an analysis of each and every revolution in history. Terrorism instils fear in the hearts of the oppressors, it brings hopes of revenge and redemption to the oppressed masses, it gives courage and self-confidence to the wavering, it spatters the spell of the superiority of the ruling class and raises the status of the subject race in the eyes of the world, because it is the most convincing proof of a nation's hunger for freedom. Here, in India, as in other countries in the past, terrorism will develop into revolution and the revolution into independence—social, political and economic.

According to this pamphlet the revolutionaries were working for the freedom of country, both openly and secretly, and in their own way. The experience of a century long and world wide struggle between the masses and the governing classes, is their guide to their goal and the methods they are following have never been known to have failed.

In this pamphlet Gandhiji was accused of misunderstanding the revolutionary psychology. In the opinion of Bhagat Singh if a man becomes a revolutionary he does not risk his life only for the sake of acquiring sympathy and appreciation from the public. He does it because reason forces him to adopt it, his conscience dictates it. It is to reason and reason alone that he bows. No amount of abuse and condemnation even if it comes from the highest of the high can turn him from his set purpose. Many a revolutionaries have laid their life for the cause regardless of the curses that the constitutional agitators rained plentifully upon them. In case one wants revolutionaries to suspend their activities the only way is to reason with them squarely. A revolutionary is the last person on earth to submit to bullying.

In the end the youth, workers, peasants and intelligentsia were expected to come forward and join the revolutionaries in their fight for freedom for establishing a new order of society in which political and economic exploitation will be an impossibility. He believed that it was due to the influence of Gandhiji that the resolution was passed by a large majority. According to Bhagat Singh in spite of the fact that the Congress was pledged to non-violence for the last ten years the supporters of the resolution indulged in abuse, called the revolutionaries 'cowards' and described their actions as 'dastardly'. In this manner their act was in contravention of the spirit of non-violence.

In this pamphlet it was mentioned that the article of Gandhiji on the cult of the bomb consisted of three things—his faith, his opinion and his arguments. As reason has little in common with faith this aspect of the article was not discussed by him. He did not believe that Gandhiji's message of non-violence was clearly understood by the masses. In his opinion an average Indian, like an average human being understood little of the fine theological nuances of '*Ahimsa*' and loving one's enemy. He raised doubts whether Gandhiji had mixed with a factory labourer or had participated actively in the social life of masses. Hence the real pulse of the nation was not known to Gandhiji. To the contrary, gospel of the revolutionaries was simple. It is, "You have an enemy, you shun him; you fight against him and if possible, kill him." From personal experiences the revolutionaries can tell that the masses were behind them. The claim of Gandhiji that he hoped to convert the foreign rulers to his way of thinking through the gospel of love and self-imposed sufferings was refuted by him. According to Bhagat Singh, up to that time Gandhiji was not able to convert any diehard English bureaucrat into a friend of India.

Bhagat Singh criticised Gandhiji for knocking the gates of the government house with beggar's bowl in his hand and dominion status on his lips. According to him, Irwin had succeeded in shattering the unity of different political parties in India which was established during the boycott of Simon Commission in India. Even the Congress today was a house divided against itself.

While praising the Congress for awakening the ignorant masses with a keen desire for freedom, Bhagat Singh wrote that in the present circumstances people have little hope from it and the revolutionaries are looking forward to the day when the masses would march arm-in-arm with the revolutionaries in their common goal of complete independence.

Bhagat Singh strongly refuted the argument of Gandhiji that violence had resulted in an increase in military expenditure of the country and consequently placed additional burden on the taxpayer. To the contrary the wars that have taken place in India since the British came here were the cause of it. The movements of *satyagraha* have also affected the finances of the government.

Bhagat Singh did not understand why Gandhiji mixed up the revolutionaries with the various constitutional reforms promulgated by the government. The revolutionaries had never claimed such reforms as their achievements. They had raised the demand for complete independence long ago and had sacrificed their life for the sake of this ideal. Such kind of sacrifices have produced a significant change in the thinking of the people.

As regards Gandhiji's contention that violence impedes the march of progress and thus directly postpones the day of freedom Bhagat Singh drew his attention to several contemporary examples where violence had led to social progress and political freedom of the people who practised it. He cited the example of Russia and Turkey. The solitary example of Afghanistan mentioned by Gandhiji was only an exception rather than a rule.

According to Bhagat Singh the great awakening among the people, during the days of non-cooperation was not due to the preaching of non-violence but it was due to the programme of direct action. In Russia, the widespread awakening among the peasants and workers came after the communists launched a programme of militant mass action, though nobody preached non-violence to them.

He attributed the failure of Gandhiji to have 'Swaraj within one year' to the policy of compromise adopted by the Congress leaders. On account of such an attitude all movements based on non-violence have ended in failure. Therefore, the sooner it is recognised that there can be no compromise between independence and slavery the better it is.

Bhagat Singh ridiculed the unfurling of the flag of independence at just one minute after mid-night of December 31, 1929. In his opinion it was not worthy of the Congress to pass a resolution for the grant of independence by the government. It should have achieved if first and declared it afterwards. Moreover, the Congress is not adopting the most effective and tried means to achieve it.

He expressed his regret over the call of Gandhiji to people to withdraw support from the revolutionaries and condemn their actions so that they may realise the futility of violence and the great harm that violent activities have done on every occasion. In his opinion Gandhiji did not and will not understand revolutionary psychology.

To think that a revolutionary will give up his ideas if public support and appreciation is withdrawn from him is a folly of highest order.

In the end Bhagat Singh exhorted the workers, peasants, youths and revolutionary intelligentsia to get together in carrying afloat the banner of freedom and help in establishing a new order of society devoid of all sorts of exploitation. He also appealed to them to have a righteous revenge on the tyrant i.e. the British which had reduced the Indians to a nation of paupers.

In these articles the views expressed by Gandhiji and Bhagat Singh are quite opposit to each other and were poles apart. The essence of Gandhiji's political ideology was non-violence. On the other hand, Bhagat Singh was in favour of violence being used against an alien government. Gandhiji believed in negotiations with the government but Bhagat Singh had no faith at all in the intentions of the British regime towards India. At the age of sixty-one Gandhiji was a matured and seasoned leader with a huge following but Bhagat Singh was hardly twenty-three years of age representing some youths of northern and eastern regions of the country who were in a haste to achieve the goal of independence by adopting all methods beyond the purview of negotiations. Every move Gandhiji made was straight, open and devoid of any kind of secrecy but the other had a zigzag way to traverse in which the maintenance of secrecy and hiding the exact identity was predominant. The elderly politician had tremendous experience of launching the *satyagraha* movement for more than two decades while Bhagat Singh had little experience for piloting a revolutionary movement at an all India level. Although they had not met each other, this was the first time that they came in confrontation with each other. On one hand, according to Gandhiji, violent activities were harming the interests of the nation. On the other hand, Bhagat Singh maintained that Gandhiji was following a policy of compromise. The masses did not understand the concept of non-violence but they were with them. He even blamed Gandhiji for destroying the unity of all parties established at the time of boycott of the Simon Commission. Looking at the circumstances which prevailed at that time, it can be said that the masses were responding to the call of Mahatma Gandhi to a great extent and it was not behind the revolutionaries as claimed by Bhagat Singh.

Here a question arises as to why Gandhiji who opposed the methods of revolutionaries made sincere efforts to save the lives of Bhagat Singh and his two comrades. The reasons are firstly, he was not in favour of anyone being sent to the gallows, secondly, he was deeply impressed by the bravery and the spirit of self-sacrifice of Bhagat Singh, and last but not the least, there was a feeling among the majority of the people that these selfless patriots should be saved from the execution. But all along Gandhiji did not lose any opportunity to express his dissatisfaction over the methods adopted by the revolutionaries.

In his last meeting with Bhagat Singh in the Central Jail, Lahore on March 23, 1931 Sardar Kishan Singh informed him about the efforts of Gandhiji and other leaders for getting the death sentence commuted to life imprisonment. Although in his heart Bhagat Singh was convinced that the alien government was not going to yield, he did not say anything to his father about it. By that time he had prepared himself completely for the last journey. But he said to Baba Chuhar Singh, a Ghadrite prisoner, that the meeting between Viceroy and Gandhi would do us no good. We three will be hanged.

The Congress policy is weak, reformatory and constitutional. They strengthen the hands of enemy. So many pacts were made previously but nothing materialised.[13]

It is rather surprising that Bhagat Singh, though opposed to Gandhism, adopted non-violent means to secure rights for the political prisoners. In Mianwali Jail he went on hunger strike and continued the same even after his transfer to Lahore Central Jail. Other political prisoners in the above jail and Borstal Jail also joined him. For all of them it was a long ordeal. After a hunger strike of eighty two days Jatindra Nath Das sacrificed his life. The hunger strikers not only earned the sympathy of the people but also the support of eminent leaders. Jawahar Lal Nehru visited them in the two jails to know about the conditions of political prisoners and provide help in getting their demands accepted by the government. In the end Bhagat Singh and his comrades were successful in getting most of their demands accepted by the government.

On the advice of Bhagat Singh the under trials of the Lahore conspiracy case adopted a non-cooperative attitude. They did not participate in the proceedings of the case. It not only delayed the hearings of the trial but also attracted publicity. Besides, while being taken or dragged to the court from the jail, they used to sing patriotic song or shouted revolutionary slogans. Before them Ram Prasad 'Bismil' and others undergoing trial in the famous Kakori conspiracy case had adopted such methods of gaining publicity and arousing patriotic feelings among the people. During the eighteen months of the trial 'Bismil' and his comrades used to sing 'Bande Matram' and others patriotic songs on their way to court from the jail. A large crowd, which was always present outside the court, also sang with them and shouted slogans. Manmath Nath Gupta has written that the revolutionaries adopted this method because they wanted publicity, without publicity they could not reach the masses, and without the coming of the masses to their aid success could not be achieved in any sphere of their activity.

In spite of being greatly impressed by the Russian Revolution Bhagat Singh was not a stone devoid of human feelings. According to him the human approach made a revolutionary different from a terrorist. In the pamphlet thrown in the Central Legislative Assembly he wrote, "We are next to none in our love for humanity. We hold human life sacred beyond words. We would sooner lay down our lives in the service of humanity than injure any one else..." In case Bhagat Singh and B.K. Dutt had decided to kill a few British bureaucrats or Indian members of the Assembly they could have easily done so but they deliberately refrained themselves from committing such an act. On this occasion their intention was to get their voice heard by the people through the corridors of the Assembly and later on through the medium of the courts.

In a message to the nation from his death cell, Bhagat Singh criticised the Congress for using the capitalists and shopkeepers for putting pressure on the British government for obtaining a few concessions. He wrote, "If the war of the country is to be fought then labourers, peasants and the common men should be brought to the forefront. They have to be organised for the war... Along with the slavery of an alien rule the peasants should also be freed from the landlords and capitalists." He further said that the slogan 'Long Live Revolution' is very sacred to them but it should be used with great discretion. He repeated his conviction expressed in the statement given before the court trying the Assembly Bomb Case, "By revolution we mean to uproot completely the present organisation and working of the society." For this purpose the revolutionaries want to capture power. For safeguarding the interests of the common man and for giving a concrete shape to their ideals the revolutionaries want the machinery of the government in their hands. They are fighting for this aim but for this they have to educate the common people. He further wrote that the aim of the Hindustan Socialist Republican Association was to establish a socialist organisation of society. Its aim was to hand the power of the government to such persons who have socialism as their goal. He again stressed that for achieving this objective it was necessary to organise peasants and labourers. In the end he wrote that by using bombs one cannot achieve anything. The throwing of bomb is not only useless but many a times it is harmful. The final aim of revolutionaries was to organise the peasants and labourers.[14]

In his speech made just before his execution, he said, "The real revolutionary armies are in the villages, in factories, the peasantry and the labourers. But our bourgeoie leaders do not and cannot dare to tackle them." The sleeping lion, once awakened from its slumber, shall become irresistible even after the achievement of what our leaders aim at. He repeated the definition of the revolution given in the statement submitted to the judge in the Assembly Bomb Case by saying that the complete overthrow of the existing social order and its replacement with the socialist order was necessary. For this purpose the revolutionaries are fighting the governmental machinery. While doing so, continued efforts to educate the masses will be made to create a favourable atmosphere for the implementation of the social programme they wanted to implement. In the end he exhorted the political prisoners to have courage, perseverance and a strong determination.[15]

The statement of Bhagat Singh submitted to the judge of the Assembly Bomb Case can be regarded as the basis of his interpretation of the word 'Revolution'. His conception of revolution necessarily implied a programme of systematic reconstruction of society on a new and better adapted basis after complete destruction of the existing state of affairs (regime) is somewhat nearer to the anarchist theory of Bakunin. Although he agreed with Ram Saran Das that while striving for the establishment of an ideal society the gradual process of evolution through the education of the masses on definite lines should be adopted. Bhagat Singh remarked that in the present circumstances the education of masses in India was not possible. Hence it was necessary to capture power and then employ all its resources to educate the masses. But till then the people should be educated about their programme for creating a favourable atmosphere for its implementation.

Bhagat Singh always maintained that merely a change in the government would not serve the objectives of the revolution. In case the foreigners were substituted by capitalists, landlords and exploiters, independence or freedom would have no meaning. As a Marxist socialist he believed in the emancipation of the proleteriat and extermination of the exploiting elements. For attaining this object he was in favour of a violent movement, if necessary. In his opinion, it was difficult for a revolution to function without any kind of bloodshed. But its aim should be to end any kind of cruelty once it was successful.

Bhagat Singh was a determined opponent of imperialism. Along with the slogan 'Long Live Revolution' he always raised the slogan of 'Down with Imperialism' (साम्राज्यवाद मुर्दाबाद). He wanted to prepare the country to fight against the British imperialism with full strength. He had very clear views about life of the people both as it existed under the British rule and as to what it should be in free India. In his opinion it was the common man who was worst hit by British rule. He was so sentimental about it that whenever he referred to some of the brutalities committed by the British, tears rolled down his cheeks. He regarded it as a shame to be ruled by a handful of foreigners for so long and submit to their economic exploitation.[16] He did not like to kill people merely because they were Englishmen or their henchmen as it would serve no useful purpose. But it was obligatory for the revolutionaries to do so in order to counteract British repression. In the statement given by him in the Assembly Bomb Case he made this clear. He also stated that he and

B.K. Dutt had no intention to kill anybody in the Assembly. They only threw live bombs on empty benches and no wanton spilling of blood was intended by them. Lord Irwin, the Viceroy, rightly acknowledged that the bombs were not directed against any individual but against the institution, i.e. the British government. However, on the other side, as a measure of retaliation Bhagat Singh took an active part in the killing of Saunders.

Bhagat Singh did not believe in the efficacy of any kind of appeal to the British authorities. According to him this kind of method generates weakness in the man who does it and in the political organization also of which he was a member. He, therefore, disagreed with Gandhiji's method of non-violence. In his opinion independence was to be wrested from unwilling hands. As the basis of the British Empire was India's slavery, how could the British follow the Gandhian method of negotiations. Moreover, freedom is not worthwhile unless millions make supreme sacrifice. According to him, for securing independence an armed insurrection on a large scale was necessary.

He did not believe in any sort of compromise with the British imperialism. According to him, the British imperialists were not paying any attention to the legitimate and just requests of the people of India. In the pamphlet thrown in the Central Assembly he wrote, "It takes a loud voice to make a deaf hear." He was never reconciled to religion being associated with revolution.

During his sojourn in Calcutta, Bhagat Singh was surprised to know from the first revolutionary he met there that he believed in the existence of God. To his surprise another Bengali revolutionary was a firm believer in God. He learnt from the Bengali revolutionaries that in the early days of their movement the revolutionaries in Bengal considered Muslims as an obstacle in the way of India's freedom. Therefore, they had an anti-Muslim feeling. Bhagat Singh was also surprised to find that several revolutionaries who faced death bravely had also a religious bent of mind. At the time of his hanging in 1908 Khudi Ram Bose had *Gita* in his hand. Ram Prasad 'Bismil' was an orthodox Arya Samajist. During imprisonment several revolutionaries expressed their belief in religion. Sachindra Nath Sanyal sang the glory of God in the first page of his book *Bandi Jebon*. Rajen Laheri, a prisoner of the Kakori Conspiracy case recited hymns from *Upanishads* and *Gita*. Four other prisoners of this case, who were sentenced to death, passed their lives in prayer. But Bhagat Singh was not in favour of mixing religion with politics. He believed in Karl Marx who said, "religion was the opium of masses". Atheism was an essential part of his political ideology. He wanted to keep his party secular.

Besides having his party to maintain its secular character, Bhagat Singh was opposed to the spirit of communalism in politics. He was very much pained to see the growth of communal organizations and the spurt of communal frenzy in the country. In 1927, he condemned communal killings as barbaric. In his opinion the communalists did not kill a person because he was guilty of a particular act but simply because the person appeared to be a Hindu, Muslim or a Sikh. In the Naujawan Bharat Sabha the youths belonging to religious communal organization were not permitted to become its members. Religion was regarded to be one's private concern and communalism was an enemy to be fought.[17]

Besides atheism a few other traits of his beliefs and character also influenced the political ideology of Bhagat Singh. Bravery, fearlessness and firm determination characterised his ideology. On no occasion he showed an element of fear or cowardice. His refusal to accept the suggestion of Chandra Shekhar Azad to escape after the throwing of bombs in the Assembly is a glaring example, not only of his spirit of self-sacrifice but also of fearlessness. For the sake of his country and the revolutionary ideals he could go to any extent. He believed that sacrifices on the altar of revolution would bring freedom and help in establishing a society in which exploitation by man against men was impossible. He lived and died cheerfully for the cause he cherished throughout his life. In the annals of the freedom struggle of the country from the British domination, his name will always be written in golden letters. He really deserves to be called as the 'prince' of martyrs.

REFERENCES

1. Transcript of an Interview with Jaidev Gupta, Nehru Memorial Museum and Library, New Delhi.
2. Jagmohan Singh and Chaman Lal (eds.), *Bhagat Singh aur Unke Sathion Ke Dastavez*, *op. cit.*, p. 23.
3. Diary of Bhagat Singh available in Nehru Museum, New Delhi.
4. Delhi Assembly Bomb Case Proceedings, Statement of Bhagat Singh and B.K. Dutt dated June 6, 1929.
5. The Revolutionary, A Paper of the Hindustan Socialist Republican Association, January, 1925.
6. Delhi Assembly Bomb Case, Statement of Bhagat Singh and B.K. Dutt, June 6, 1929.
7. Sandhu, Virendra, *Patra Atir Dastavej*, Delhi, 1974, pp. 46-49.
8. *Ibid.*, pp. 81-88.
9. *Ibid.*
10. *Ibid.*
11. Jagmohan Singh and Chaman Lal, Documents of Bhagat Singh and his Associates, *op. cit.*, p. 26.
12. Sandhu, Virendra, *Patra Atir Dastavej, op. cit.*, pp. 46-49. However, according to Bipan Chandra it was drafted by Bhagwati Charan Vohra in consultation with Chandra Shekhar Azad and Yashpal. See Bipan Chandra and others, *India's Struggle for Independence, op. cit.*, p. 255.
13. Bakshi, S.R., *The Struggle for Independence, op. cit.*, pp. 181-182.
14. Sandhu, Virendra, *Shaheed Bhagat Singh , op. cit.*, pp. 88-92.
15. Transcript of an Interview with Durga Das Khanna, Nehru Memorial Museum and Library, New Delhi.
16. *Ibid.*
17. Chandra, Bipan, and others, *India's Struggle for Independence, op. cit.*, p. 257.

4

IDEOLOGY AND PERCEPTION

Bhagat Singh represented the popular upsurge of his time. He was brave, calm and fearless, and considered the attainment of freedom and thereafter the establishment of a socialist republic of workers and peasants in India the sole aim of his life.

Bhagat Singh was a first rate patriot who thought of nothing else than his country. Whatever he did, he did for the cause of the motherland. He worked day and night for attainment of India's freedom through revolutionary methods. Speaking on the move for adjournment in the Central Legislative Assembly, Pandit Moti Lal Nehru remarked that those youngmen (Bhagat Singh and others) were "devoted and high souled persons".[1] Commenting on the patriotic work of Bhagat Singh, Pandit Madan Mohan Malaviya said in the Central Legislative Assembly, while supporting the adjournment motion tabled by Pandit Moti Lal Nehru, that "Bhagat Singh and his comrades are not ordinary criminals. However much one may condemn any act of violence of which they might be proved guilty, they are not prompted by any sordid selfish motive. They are ...persons who are inspired by a high sense of patriotism and a burning desire for the freedom of their country. They are men of high ideals and possess a high sense of national self respect".[2]

Patriotism was indeed a ruling passion with Bhagat Singh. No personal or creative need could deflect him from his chosen path. Accordingly, when his people pressed upon him the necessity of marriage, he rejected the proposal outright. At that time, he was still a student of B.A. class. From the letter which he wrote to his father, one can measure the depth of his patriotic sentiments :

"Respected Father,

This is not the time for marriage. The country is calling me. I have taken an oath to serve the nation bodily, mentally and monetarily (*tan, man te dhan nal*). Moreover, it is not a new thing for us. Our whole family is full of patriotism. After two or three years of my birth in 1910, uncle Swarn Singh died in jail. Uncle Ajit Singh is living the life of an exile in foreign countries. You have also suffered a lot in jail. I am only following in your footprints and, thus, dare to do this. You will kindly not tie me down in matrimony, but give your blessings that I should be successful in my mission." His devotion to the nation is evident from another letter which he wrote to his father declining and condemning his move for marriage.

Respected Father, I am astonished to receive and read the contents of your letter. When you, who is a staunch patriot and brave personality, can be influenced by such trifles, then what will happen to an ordinary man ?

You are only caring for *dadi* (my grandmother), but in how much trouble is our mother of 33 crores, the Bharat Mata ? We will have to sacrifice everything for the sake of her troubles.

I know that I shall be compelled to get myself married. Therefore I am leaving this place and going somewhere else."[3]

Commenting on the partiotism and sacrifice of Bhagat Singh, Dr. Karan Singh writes, "Among those who laid down their lives so that India could become free, one of the bravest and most dramatic figures was Bhagat Singh. Coming as he did from the Punjab, with its tradition of virility and courage, Bhagat Singh by a supreme act of sacrifice became a symbol of the spirit of revolt that swept the younger generations in India at that time. His story has become a legend, and his name has become synonymous with patriotism and sacrifice in the cause of freedom".[4] Service of the motherland and its people was in fact the religion of Bhagat Singh. In his last message to his younger brothers, he said, "Brothers don't lose heart. After I am dead, don't give up the service of the country and of the people. Sacrifice yourselves at the altar of the nation."[5]

Views about on Nationalism and Internationalism: One of the objects of the *Naujawan Bharat Sabha* was to infuse a spirit of patriotism in the hearts of the youth of the country in order to establish a united Indian nation. He was prepared to make any sacrifice for India's national independence. When he went to Kanpur in the year 1924 and came in contact with Ganesh Shanker Vidyarathi, who told him about the sufferings and sacrifices which a lover of the nation had to undergo like the moth who loves the burning candle, Bhagat Singh replied, "I have come here having taken a vow to do and die for my country's liberty". His deep rooted devotion to the cause of Indian nationalism is also evident from the following message he gave to his friends before parting. He said, "Friends, I tell you today that if my marriage takes place in India when she is under subjection, my bride shall be only death. The *barat* will take the form of a funeral procession and the *baraties* will be the martyrs in the cause of India". In fact, service of the nation was adopted by Bhagat Singh as a sort of religion when he was six years old. On having been asked by Anand Kishore Mehta, a friend of Kishan Singh, 'What is your religion, my boy?' Bhagat Singh replied, 'Service to my nation (*desh bhakti*)'.

No doubt, Bhagat Singh was a staunch nationalist but this does not mean that he stood for narrow and intense nationalism. He was also an internationalist and believed in internationalism and humanism. He held that one of the aims of the *Naujawan Bharat Sabha* should be "expression of sympathies and the rendering of help to all economic, industrial and social movements free from communal sentiments, going on in any part of the world". In his statement in court, Bhagat Singh emphatically declared, "We are next to none in our love for humanity, far from having any malice against any individual, we hold human life sacred beyond words".[6]

Views about Secularism: Bhagat Singh had a secular outlook from the very beginning and believed that communal sentiments must not overpower a patriot in the service of the nation and humanity. The *Naujawan Bharat Sabha* was based on secularism. Members of all castes and communities could join the sabha. Before admission to the Sabha, each member was made to sign a pledge that he would place the interests of his country above those of his community. One of the political aims of the Sabha further prescribed that each member was "to express sympathy with and to assist the economic, industrial and social movements, which being free from communal sentiments, are intended to take us nearer to our ideal, namely the establishment of completely independent republic of labourers and peasants".

Bhagat Singh did not adopt any type of religious or communal salutation or mode of greeting. In his letters written to his father, brothers and friends, he made use of all forms of greetings; for example in his letter dated 26th April, 1929, addressed to his father, Kishan Singh, he used the salutation 'Bande Matram'. In another letter written to his younger brother, Kulbir Singh, sent on the 6th of September, 1930, from Lahore Central Jail, he used the greetings of 'Sat Sri Akal'. In his last letter to his brother, Kultar Singh, dated 3rd March, 1931, he wrote at the end 'Namaste'. Indeed, he was above communalism and was secular from the core of his heart.

Moreover, the Hindustan Republican Association, was secular in character. Every lover of freedom was entitled to become its member, irrespective of caste, creed or community. People from United Provinces, Punjab, Rajasthan, Central Provinces, Bengal and other provinces were members. They were enjoined to rise above parochialism, cultivate a national and socialist sentiment and join the organization not as members of any particular sect or community, but only as Indians. All the members worked, sat and ate together. Next to the Ghadrites, Bhagat Singh and his comrades were the pioneers of the principle of secularism in India, which now forms a part of the Preamble to the Indian constitution.

***Serious Student of History*:** Bhagat Singh was not an ordinary freedom fighter but a serious student of history and revolutions. In the words of Gopal Thakur, "Bhagat Singh had no boyish illusions that the explosion would bring about a revolution or even that it would be a step towards revolution. The travails of his subjugated motherland created a rebellious impulse in him. The throwing of bombs was to be an expression of his 'anguish' over the 'heart rending agony' of his fellow beings".[7] With a deep and burning sense of injustice prevailing in the social set-up in the country at the time, he told the court, when he was on trial, "We are neither perpetrators of dastardly outrages and therefore a disgrace to the country as the pseudo-socialist Dewan Chaman Lal is reported to have described us, nor were we 'lunatics' as, the *Tribune* of Lahore and some others would have it believed. We humbly claim to be no more than serious students of history and conditions of our country and human aspirations and we despise hypocrisy. Our practical protest was against the institution which since its birth has eminently displayed not only its worthlessness but its far reaching power for mischief. The more we have pondered, the more deeply we have been convinced that it exists only to demonstrate to the world the Indian humiliation and helplessness, and it symbolises the overriding domination of an irresponsible and autocratic rule.

"Time and again, the national demand has been pressed by the representatives of the people, only to find the waste paper basket as its final destination. Solemn resolutions passed by the house have been contemptuously trampled underfoot, on the floor of the so-called Indian parliament. Resolutions regarding the repeal of repressive and arbitrary measures have been treated with sublime contempt and government measures and proposals rejected as unacceptable by the elected members have been restored by a stroke of pen.

"In brief, in spite of earnest endeavour, we have utterly failed to find any justification for the existence of an institution which despite all pomp and splendour, organised with the hard earned money of the sweating millions of India, is only a hollow show and a mischievous make-believe. And have we failed to comprehend the mentality of public leaders who help to squander public time and money on so manifestly a stage managed exhibition of India's subjection ? We have been ruminating upon all this as also upon the wholesale arrests of the leaders of the labour movement when the introduction of the Trades Disputes Bill brought us into the assembly to watch its progress and the course of the debate only served to confirm our conviction that the labouring millions of India had nothing to expect from an institution that stood as a menacing monument to the strangling power of the exploiters and the serfdom of the helpless labourers.

"Finally an insult which we considered inhuman and barbarous was hurled on the devoted heads of the representatives of the entire country and the starving and struggling millions were deprived of their primary right and the sole means of their economic welfare.

None who has felt like us for the dumb-driven drugdes of labourers could possibly witness this spectacle with equanimity, none whose heart bleeds for those who have given their life blood in silence to the building up of the economic structure of the exploiters, of whom the government happen to be the biggest in the country, could repress the cry of the soul agonising anguish which is so ruthlessly wrung out of our hearts.

"Consequently, bearing in our minds the words of the late Mr. S.R. Das, once the law member of the Governor General's Executive Council, which appeared in the famous letter he had addressed to his son to the effect that the bomb was necessary to awaken England from her dreams, we dropped bombs on the floor of the assembly chamber to register our protest on behalf of those who had no other means left to give expression to their heart rending agony. Our sole purpose was 'to make the deaf hear' and to give the heedless timely warning."[8]

***India of Bhagat Singh's Dream*:** Bhagat Singh aimed at building India on new lines. It is evident from the following extract of the reply which he gave to his mother, who joined his father in urging marriage. She wanted him to lead his life peacefully and not spend it worrying day and night over an abstract ideal, such as independence. He responded:

"The country is in slavery. Slavery has given rise to hunger and starvation. Owing to the existence of British imperialism, Indians are not regarded as human beings. To win the country's freedom, we have to embark upon big projects. The country is to be made free and after freedom a new India is to be built up. Palatial buildings are to be constructed in place of huts. The naked and the hungry are to be provided with food and clothing". Moreover, Bhagat Singh wanted the new India to be built on principles of complete, national sovereignty, independence, the organization of workers and peasants, and, socialism. One of the political aims of the *Naujawan Bharat Sabha* was the establishment of a completely independent republic of the labourers and peasants of the whole of India".[9]

Critique of Congress Leadership: Bhagat Singh was a severe critic of Congress leadership and its methodology and ideology. Their acceptance of the theory that under imperialism and capitalism a small privileged class could thrive on the merciless exploitation of the vast mass of common people, gave Bhagat Singh a keen insight into the political nature of Congress leadership. Fear of a compromise between the Congress and the British government found expression, even in 1928, in the *Naujawan Bharat Sabha* manifesto, mainly drafted by Bhagat Singh and Bhagwati Charan. It said :

"Along with the advent of the 20th century, the British bureaucracy has adopted quite a new policy towards India. They are drawing our bourgeoisie and petite-bourgeoisie into their fold by adopting the policy of concession. Their cause is being made common. The progressive investment of British capital in India will inevitably lead to that end. In the very near future we will find that class and their great leaders having thrown their lot with the foreign rulers. Some round table conference or any such body will end in a compromise between the two. They will not help the masses any more".[10]

Bhagat Singh was not opposed to compromises as such. He accepted them as expedients in all political bargaining. He held, however, that they should not be anything more than respites in the struggle, meant only to help increase the tempo thereafter. A negotiated settlement or a compromise should not be a substitute for the achievement of the main goal. His characterisation of the Congress from this point of view is clearer than the one given in the *Naujawan Bharat Sabha's* manifesto. In his message from prison, he said :

"The Congress seeks to gain some power by getting business-men and capitalists to exercise pressure on England. But in this does not lie salvation for the millions of workers and peasants of India. The nation can wage a successful struggle only on the strength of organised workers, kisans and the common people. The leaders (Congress) have done nothing to mobilise these sections of the population. The peasants have to liberate themselves not only from foreign yoke but also from the yoke of landlords and capitalists. This is not the goal of the Congress. It cannot be said to be standing for complete revolution. Get some concessions for India's wealthy class through economic pressure on government, that is what the Congress wants. In my opinion, therefore, the Congress movement will liquidate itself in a compromise or end up in failure".[11]

Views about Method of Struggle: Bhagat Singh was opposed to the method of petitions and prayers. He declared in his statement before the court that tried him that it needed 'a louder voice to make the deaf hear', and that 'the bomb was necessary to awaken England from her dreams'. He believed in individual suffering for the common good. He said that "the sacrifice of a few individuals at the altar of the great revolution that will bring freedom to all, rendering exploitation of man by man impossible, is inevitable". But this does not mean that he was a blood thirsty person. He said in his statement in the court and in the leaflet thrown in the Assembly: "We attach great sanctity to human life, we regard human life as sacred as anyone else could. We would sooner lay down our life in the service of humanity than injure any one else." He even declared that "India's struggle will continue as long as a handful of men in power continue to exploit the labour of the common people for their own ends. Whether the exploiters are Indian or British or both in alliance, nothing else shall halt the struggle."[12]

Commenting on the method of struggle in which Bhagat Singh and his comrades believed, Ajoy Ghosh writes, "As for the most important question, it was the question in what manner the fight for freedom and socialism was to be waged. Armed action by individuals and groups was, however, to remain our main task. Nothing else, we held, could smash constitutional illusions, nothing else could free the country from the grip in which fear held it. When the stagnant calm was broken by a series of hammer blows delivered by us at selected points and on suitable occasions against the most hated officials of the government, and mass movement unleashed, we would link ourselves with that movement, act as its armed detachment and give it social direction."[13]

Views about Violence: Violence meaning physical force applied for committing injustice was not how Bhagat Singh interpreted the word. On the other hand, he said, "What generally goes by the name of non-violence, is in reality the theory of soul- force, used for achieving personal and national rights through self- suffering. When a revolutionary believes certain things to be right, he asks for them, pleads for them, argues for them, wills to attain them with all the soul-force at his command. He stands the greatest amount of suffering for their sake, is always prepared to make the highest sacrifice for their attainment and also backs his efforts with all the physical force, he is capable of. So he adds physical force to the soul-force".

Bhagat Singh believes that the politics of "non-violence has proved its futility to the rising generation beyond shadow of doubt". He further says that "force when aggressively used is violence, and is, therefore, morally unjustified. But when it is used in furtherance of a legitimate cause, it has its moral justification. Elimination of force at all costs is Utopian, and the new movement which has arisen in the country and of which we have given a warning, is inspired by the ideals which guided Guru Gobind Singh, Shivaji, Kemal Pasha and Riza Khan, Washington and Garibaldi, Lafayette and Lenin."[14]

***Concept of Terrorism*:** Bhagat Singh had a very clear concept of terrorism. He held the view that terrorism cannot be divorced from the mass action necessary to overthrow an unjust order. In *The Philosophy of the Bomb*, Bhagat Singh and his comrades gave a clear definition of terrorism. They declared : "Terrorism is a phase, a necessary inevitable phase, in the revolution. Terrorism is not the complete revolution and the revolution is not complete without terrorism. This thesis can be supported by analysis of any and every revolution in history.... Terrorism shatters the spell of superiority of the ruling class and raises the status of the subject race in the eyes of the world because it is the most convincing proof of a nation's hunger for freedom. Here, in India, as in other countries in the past, terrorism will develop into revolution, and revolution into independence—social, political, economic."[15] From this it is evident that terrorist acts, in the opinion of Bhagat Singh, were a historically necessary forerunner to a takeover by the masses and the establishment of government for the masses. But the revolution by the masses could only come by the organization of workers and peasants. In theory, the vital importance of this work was not lost upon Bhagat Singh and his comrades inspite of their immediate belief in armed action as the immediate programme before them".[16]

***Views about Armed Action and Mass Movement*:** As mentioned earlier, Bhagat Singh believed in 'individual armed action' but, at the same time, he knew the urgent necessity of a mass movement, with the help of which alone could real change be brought about. He considered immediate armed action by some individuals or groups as an immediate task for the revolutionaries, while for the communities, the foremost task was to organise the workers and peasants. Although he differed from the communists with regard to the sequence of actions, he did not reject the necessity of a mass movement against British rule. For him, individual acts of terrorism were only a means of rousing the people to action. Published before the assembly bomb incident, the *Naujawan Bharat Sabha's* manifesto was primarily an impassioned appeal to Indian youth to make sacrifices for the cause of the nation. It declared, "The young can bear the most inhuman tortures smilingly and face death without hesitation because the whole history of human progress is written with the blood of young men and women."

The manifesto further declared, "Do you not know the wonders worked by the Turks ? Do you not daily read what Chinese are doing ? Was it not the young Russians, who sacrificed their lives for Russia's emancipation? Through the last century, hundreds and thousands of them were exiled to Siberia for the mere distribution of socialist pamphlets or, like Dostoevesky, for merely belonging to a socialist debating society. Again and again they faced the storm of oppression. But they did not lose courage.

The manifesto did not fail to mention the necessity of work among the people. It said, "The future programme of preparing the country will begin with the motto : 'Revolution by the masses and for the masses'. In other words, Swaraj for the 98 per cent, Swaraj not only attained by the masses but also for the masses. Without going into details, we can safely assert that to achieve our object, thousands of our most brilliant young men, like the Russian youth, will have to pass their precious lives in villages and make the people understand what the Indian revolution would really mean. A revolutionary does not necessarily mean a man of bombs and revolvers".[17]

Concept of Independence: It was not only on the use of methods that Bhagat Singh differed from the Congress and Gandhi. A more substantial difference was the concept of independence.

The Congress was leading a movement against the British Rule. But to Bhagat Singh, that movement would not release India's toiling people from all economic fetters clamped on them by imperialism. For him, the slogans "Long Live Revolution" and "Down with Imperialism" were inseparably linked with the complete economic emancipation of Indian peasants and workers. This approach to independence from the British rule was implicit in the adoption by him and his comrades of a 'Socialist Republic' as their objective as early as 1928. His trial gave him an opportunity to describe his ideal in detail.

***Views about Socialism*:** Bhagat Singh had an unshakable faith in socialism and was an ardent advocate of what can be termed as scientific socialism. He was a true scientific socialist and had deep love for the working class. This is evident from the objectives which he kept before the *Naujawan Bharat Sabha*. These were to "organise the labourers and the peasants, and establish a completely independent republic of labourers and peasants of the whole India".[18] In the Delhi Court, he said that "the present order of things, which is based on manifest injustice must change. The producers or labourers in spite of being the most necessary element of society, are robbed by their exploiters of the fruits of their labour and deprived of their elementary rights. The peasant who grows corn for all, starves with his family; the weaver, who supplies the world market with textile fabrics, has not enough to cover his own and his children's bodies; masons, smiths and carpenters who raise magnificent palaces, live like pariahs in slums. The capitalists and exploiters, the parasites of society, squander millions, on their whims. These terrible inequalities and forced disparity of chances are bound to lead to chaos. This state of affairs cannot last long, and it is obvious, that the present order of society in merry making is on the brink of a volcano.

"The whole edifice of this civilization, if not saved in time, shall crumble. A radical change, therefore, is necessary and it is the duty of those who realise it to reorganize society on the socialistic basis. Unless this thing is done and the exploitation of man by man and of nations by nations is brought to an end, sufferings and carnage with which humanity is threatened today cannot be prevented. All talk of ending wars and ushering in an era of universal peace is undisguised hypocrisy". Bhagat Singh regarded "the workers as the real sustainer of society." He held the view that "the sovereignty of the people was the ultimate destiny of the working class", for which he was prepared to do anything including laying down his life.

From this, it is clear that Bhagat Singh wanted to "establish a new social order based on justice, ensuring their due both to the workers and to the peasants. To achieve this, capture of state power is necessary." He said, "The state apparatus is now in the hands of the privileged classes. The protection of the interests of the masses, the translation of our ideal into reality, that is, the laying down of the foundation of society in accordance with the principles of Karl Marx, demand our seizure of this apparatus". This he wanted to do through revolution. To him, revolution was an "inalienable right of mankind".[19]

Writing about this, Chaman Lal in his book entitled, '*Martyrs of India*' observes, "In a very long statement, which has now become a historic document, they (Bhagat Singh and B.K. Dutt) said that the immediate aim of the revolutionary party was to bring about the dictatorship of the proletariat and the peasantry in India. This statement was hailed as a great contribution to political thought. Their ideas received particular attention, because their words did not emanate from the chair of a professor but came forth from the mouths of the revolutionaries who were ready to lay down their lives for the fruition of their ideas."[20]

Hailing Bhagat Singh's contribution to socialism, Bejoy Kumar Sinha wrote, "Bhagat Singh laid down his life for the great socialist cause. When he met his death in a British prison, it was not as a fighter against a people who happened to belong to an alien race, but as against a hated imperialist system that cruelly exploited his countrymen. It would be wrong to recognize him only as a great patriot, for he was greater still as the ideal representative of a new era in our national struggle—an era that ushered in new ideals and ideas in our political movement. His eventful revolutionary life dramatically symbolised the rising temper of the fighting Indian people and of their favour."[21] Before his ascending the gallows, in his last message to his comrades, Bhagat Singh said, "Friends, this meeting and parting is final. We may not be able to meet again. When you complete your imprisonment and go home, don't get involved in worldly affairs. Unless you drive out the British from India and establish a socialist democracy, don't sit comfortably. This is my last message to you."[22]

Views about Revolution: Bhagat Singh's views about revolution and socialism were closely connected. Having been asked by the court what he meant by revolution, he declared, "Revolution does not necessarily involve sanguinary strife, nor is there any place in it for individual vendetta. It is not the cult of the bomb and the pistol. By revolution, we mean that the present order of things, which is based on manifest injustice, must change". Further elaborating his views, he continues, "By revolution we mean the ultimate establishment of an order of society in which the sovereignty of the proletariat should be recognized and a world federation should redeem humanity from the bondage of capitalism and the misery of imperial wars."[23] He declared further in the court: "This is our ideal, and with this ideology as our inspiration, we have given a fair and loud enough warning.

"If, however, it goes unheeded and the present system of government continues to be an impediment in the way of natural forces that are swelling up, a grim struggle will ensue, involving the overthrow of all obstacles, and the establishment of the dictatorship of the proletariat to pave the way for the consummation of the ideal of revolution. Revolution is an inalienable right of mankind. Freedom is an inalienable right of mankind. Freedom is an imperishable birth right of all. Labour is the real sustainer of society. The sovereignty of the people is the ultimate destiny of the workers.

"For these ideals, and for this faith we shall welcome any suffering to which we may be condemned. At the altar of this revolution we have brought our youth as an incense, for no sacrifice is too great for so magnificent a cause. We are content, we await the advent of Revolution. Long Live the Revolution."[24]

Even the leaflet which he threw into the Assembly Chamber contained reference to the French revolution and ended up with the slogans: "Down, down with imperialism"; "Long Live the Revolution". To him, the "definite revolution he aimed at was a continuous one which had been in progress in other parts of the world and had to be brought about in India.'[25] Bhagat Singh introduced the slogan, 'Long Live the Revolution' which subsequently became the 'battle cry' for all the militant elements in the country, irrespective of party creed.

Advocate of a Classless Socialist State: Bhagat Singh was an advocate of classless socialist state. He wanted a new type of social order, where there could be no exploitation of man by man. Elaborating his view on revolution, Bhagat Singh spoke of this ideal in this way : "The revolution, will ring the death knell of capitalism and class distinctions and privileges. It will bring joy and prosperity to the starving millions who are sweating today under the terrible yoke of both foreign and Indian exploitation. It will bring the nation into its own. It will give birth to a new state, a new social order. Above all it will establish the dictatorship of the proletariat and will for ever banish social parasites from the seat of political power." Commenting on this aspect of Bhagat Singh's thinking, Shiv Varma writes, "He (Bhagat Singh) had come very near Marxism towards the end of his life."[26] Bhagat Singh himself wrote in a letter to Sukhdev, "You and I may not live but our people will survive. The cause of Marxism and Communism is sure to win."[27]

It may be mentioned here that Bhagat Singh and his comrades were pioneers in calling for the establishment of a socialist republic in India. It was in September 1928, that the Hindustan Republican Association held its meeting at Feroze Shah Kotla, Delhi. It was at this meeing that Bhagat Singh suggested the word, 'Socialist' to be added to the nomenclature of the Association, as indicative of the type of system India would have after the revolution. His proposal was accepted and the erstwhile Hindustan Republican Association came to be known as the Hindustan Socialist Republican Association. From then on, the attainment of a Socialist Republic in India became the goal of the association. It can thus be said that Bhagat Singh and his comrades paved the way for the adoption of socialism as one of the principles of the Indian political system, which now forms a part of the Preamble to the Constitution of India.

***Views about the Constitution of Independent India*:** As a political thinker, Bhagat Singh was a realist in the sense that he and his comrades held that the constitution of the new independent India should be framed by the representatives of the people elected on the basis of universal adult franchise. In the objectives of Hindustan Republican Association, it was clearly laid down that "the final form of the constitution of the republic shall be framed and declared by the representatives of the people at a time when they will be in a position to enforce their decisions." And, "the basic principle of the republic shall be universal suffrage and the abolition of all systems which make any kind of exploitation of man by man possible".[28] This implies that the new Constitution of India which Bhagat Singh had envisaged was to be based on elimination of capitalism and casteism, and the establishment of secularism, socialism, and the sovereignty of the working class.

Scientific Socialist and True Revolutionary: From the foregoing account of the entire ideology of Bhagat Singh, it is clear that he was a believer in and advocate of scientific socialism. He was a socialist from the core of his heart. Therefore to regard Bhagat Singh only as a freedom fighter is to do injustice to the work of this great son of India. As a scientific socialist, he correctly visualised the 'future course of political development in India. It was he who believed among other things, that the task of leading and carrying forward the freedom struggle would fall to the workers and peasants, to the common people led by the proletariat and a party based on the masses and relying on the teachings of Marx: These points clearly emerge from the letters and documents written by Bhagat Singh, especially the manifesto of the *Navjawan Bharat Sabha* of which he was the main draftsman, and the statement which he gave in court. His socialist leanings are evident from another slogan which the Hindustan Socialist Republican Association had adopted which Bhagat Singh often repeated and shouted in the court. It was "God Soviet Guide Our Way". It implies that Bhagat Singh regarded the working class as his God.

As a genuine revolutionary, Bhagat Singh believed that the end justifies the means. Throughout his short political career, he neither cared for himself nor tried to save himself at a time when duty indicated otherwise. He was the first revolutionary whose belief in terrorism was based on an intelligent study of the revolutionary movements of the world. Even Justice Ford said: "Bhagat Singh is a sincere revolutionary and believes that the world can be improved by destroying the social structure, as it now stands". As a true revolutionary, he condemned suicide just to end one's own agonies. This can be corroborated by an extract from Bhagat Singh's historic letter to Sukhdev who was thinking of committing suicide to put an end to all troubles and torture in the jail.

Bhagat Singh wrote, "Suicide is a hated crime. It is a clear act of cowardice. Not to speak of being called revolutionary, nobody will declare this act as honourable. This death (through hanging) shall be beautiful. But to commit suicide just to avoid torture and kill one's self is cowardice. I tell you, troubles will make a man complete and perfect."[29]

Gulab Singh, formerly Chief Parliamentary Secretary of the composite Punjab, in his tribute to Bhagat Singh observed, "Bhagat Singh was a revolutionary Sardar. He raised the slogan 'long live revolution' at a time when the shouting of this slogan fetched the death-sentence."[30]

In fact Bhagat Singh was the first revolutionary whose belief in terrorism was based on an intelligent study of the revolutionary movements of the world. He was a very well read youngman, who had made an extensive study of many revolutionary writers. His statement in the court will ever rank as one of the most important revolutionary documents in the world.

Bhagat Singh's role, aim and ideology of life was best summed up by Partap Singh Kairon : "Bhagat Singh has become a symbol for our young men. The way he fought against imperialism, underwent torture and laid down his life is an everlasting example of individual bravery and unflinching faith.

"The memory of Sardar Bhagat Singh and his comrades will be revered forever because of the role they played after the martyrdom of Lajpat Rai. The fact which gives him a place in the history of the freedom struggle is his ideal of a socialist republic, for which he fought. Bhagat Singh had three aims in life: to live, to fight and to die",[31] obviously for his motherland.

REFERENCES

1. *Legislative Assembly Debates*, Vol. I, IV, No. 9, p. 1.
2. *Ibid*, pp. 3-4.
3. Deol, G.S., *Shaheed Bhagat Singh, op. cit*., pp. 17-18.
4. *Ibid*. Foreword by Dr. Karan Singh, p. x.
5. *Ibid*, p. 82.
6. Statement of Bhagat Singh in the Delhi Court.
7. Thakur, Gopal, *Bhagat Singh, The Man and His Ideas* (New Delhi: Anmol Publishers), 1989, p. 33.
8. Extracts from the Statement of Bhagat Singh given in the Delhi Court.
9. Home Department (Pol.), Government of India, 1930, and K.W. p. 1.
10. Quoted from Gopal Thakur's Book, *Bhagat Singh The Man and His Ideas, op. cit*., p. 42.
11. As quoted by Gopal Thakur in his book on *Bhagat Singh, The Man and His Ideas:* (New Delhi: Anmol Publishishers), 1990, p. 43.
12. *Ibid*, 43.
13. Ghosh, Ajoy, *Bhagat Singh and His Comrades* (Bombay: Allied Publishers), 1945, p. 4.
14. Statement given by Bhagat Singh in the Delhi Court.
15. The Philosophy of the Bomb: As quoted by Gopal Thakur in his book on *Bhagat Singh, The Man and His Ideas*, op. cit., p. 45.
16. *Ibid*, p. 46.
17. Quoted from Gopal Thakur's Book, *Bhagat Singh, The Man and His Ideas, op. cit*., pp. 44-45.
18. Home Department (Political) 1930. File No. 130, p. 1.
19. Statement of Bhagat Singh in the Delhi Court.
20. Chaman Lal, Martyrs of India (Kanpur: Chaman Lal), 1957, pp. 97-98.
21. *The Tribune*, dated 23rd March, 1967 Article on Sardar Bhagat Singh by Bejoy Kumar Sinha.
22. Quoted from Dr. G.S. Deol's book, *Biography of Bhagat Singh*, p. 76.
23. Statement of Bhagat Singh in the Delhi Court.
24. Extract from the statement of Bhagat Singh.
25. Asaf Ali, Article on Bhagat Singh: An Obsolescent, published in the *Commonweal* of March 23, 1949, p. 3.
26. Shiv Varma, Foreword to Gopal Thakur's book on Bhagat Singh, *op. cit*., p. 2.
27. *Ibid*.
28. Home Department (Pol.) Government of India, File No. 375 (1925) Papers regarding Hindustan Republican Association.
29. Quoted from the Daily Preet Jullundur, Bhagat Singh Number 1, 23rd of March, 1949, p. 19.
30. Quoted from an article on Bhagat Singh by Sardar Gulab Singh.
31. Tarlok Singh, Hindustan da Quami Shaheed Bhagat Singh, Amritsar Tribute by Partap Singh Kairon (Former Chief Minister of Punjab) p. 63.

5

CONGRESS, GANDHI AND BHAGAT SINGH (DIFFERENT PERSPECTIVES)

The correspondence of Sachindranath Sanyal, the right hand man of Rash Bihari Bose, the mastermind behind the bomb thrown at Lord Harding in 1912, with Mahatma Gandhi has been reproduced in this chapter in total. This itself shows the difference of approach between revolutionaries and Mahatma Gandhi. A four-page leaflet entitled *The Revolutionary* was published and secretly distributed all over India from Peshawar to Rangoon. This wide distribution was meant to give the police and the public impression that the party had a big organisation. Thus the leaflet was a tremendous success. It revealed to a very wide public that the revolutionary party had certain very lofty social ideals. It was not fumbling and knew what it was after. The leaflet began with the quotation "Chaos is necessary to the birth of a new star".[1] Following the line of the yellow paper, it mentioned Soviet Russia as well as the seers of ancient India. The manifesto refused to countenance the charge that it was a terrorist party. It said, it had no faith in terrorism, but in case the party was forced, it would enter into a desperate campaign of terrorism in which the life of every Englishman and his Indian lackey in India would be made impossible.

It was followed by a Bengali leaflet written by Sanyal under his signature styled as "Deshbashir Prati Nivedan", i.e. an appeal to my countrymen. All these together reveal that the HRA wanted to cut itself as under from the old revolutionary ideology of religious nationalism, but under a leader of Sanyal's eminence and calibre, it could not make much progress. At the same time Sanyal was not so blind as not to see the possibilities of the new ideas that were coming from Russia and fast infecting the younger revolutionaries, but in a self-complacent manner he fondly liked to believe that these new values and ideas and much more were included in the ideas of the seers. He was nearer to Vivekananda and Aurobindo than to Marx and Lenin.

Gandhi swore by spiritual values and so did Sanyal, but they were poles apart in theory as well as in practice. This was revealed by the correspondence they had sometime in 1925, when Gandhi after release had taken up the editorship of *Young India*. Indeed the revivified revolutionary party henceforth was forced to fight a battle on two fronts all the time. Gandhi, in order to show that he made no common cause with the revolutionaries, condemned them in season and out of season. Whenever there was an overt act, he took the opportunity to bitterly condemn them. In this his lieutenants did not always see eye to eye with him and there was open controversy between him and C.R. Das on Gopimohan Saha. Saha shot an European. He wanted to shoot Charles Tegart, the notorious police chief, instead he shot one Mr. Dey. He expressed his sorrow that he had shot the wrong man. He was hanged. He took his sentence bravely and went to the gallows laughing. C.R. Das extolled the bravery of Saha. Not only that, he went to the extent of getting a resolution passed in the Sirajgunj Bengal provincial conference, praising Saha. In reality C.R. Das became the vehicle of public opinion in Bengal. Bengal had never taken non-violence seriously, although Bengal sent more people to prison during the Non-cooperation Movement than probably the rest of India put together.

The Sirajgunj resolution was too much for Gandhi. He came out with an open denunciation. A bitter controversy followed. Ultimately Gandhi was able to scotch the resolution by getting another resolution passed in an all-India get-together of the Congress modifying it. But the modification also was praise to a certain extent. This did not win over the Bengal youth, it only aroused their ire.

Sanyal wrote the first letter anonymously to Gandhi. Rajendra Lahiri, later on hanged, brought the letter from Allahabad. "I was allowed to read it and I was asked to post it. My handwriting appeared on the envelope. The idea was perhaps that in case there was a police enquiry Sanyal should not be arrested. Gandhi published the letter as well as his reply. Sanyal did not give his real name in the letter, but I do not remember the name he used on this occasion."[2] Here is the whole text of the article as published in *Young India.*

A REVOLUTIONARY'S DEFENCE

A Correspondent, who has given his name but not his address, has sent me what he calls "an open letter". It is a letter in reply to my remarks on the revolutionary movement in my address to the Belgaum Congress. The letter breathes love of the country, fervour and a spirit of self-sacrifice. It is moreover written under a sense of wrong, said to have been done by me to the revolutionaries. I therefore gladly print the letter without the name. The address of the writer is not given. The following is the unchanged full text of the letter:

"I think it my duty to remind you of the promise you made some time back that you would retire from the political field at the time when the revolutionaries will once more emerge from their silence and enter into the Indian political arena. The experiment with the Non-violent Non-cooperation Movement is now over. You wanted one complete year for your experiment, but the experiment lasted at least four complete years, if not five, and still do you mean to say that the experiment was not tried long enough?

"You are one of the greatest of personalities in the present age and under your direct guidance and inspiration, your programme was actually taken up for some reason or other, by the best men in the land. Thousands of young men, the flower of the youth of our country, embraced your cult with all the enthusiasm they could gather. Practically the whole nation responded to your call. We can safely say that the response was phenomenal if not miraculous. What more could you want? Sacrifice and sincerity on the part of your followers were not wanting; the most selfish of professional men gave up their professions, young men of the country renounced all their worldly prospects and joined the forces under your banner;

hundreds of families were rendered destitute for want of pecuniary income. Money was not wanting. You wanted one crore of rupees and you got more than you wanted. In fact I perhaps be not far from the truth if I say that the response to your call was more than you yourself expected. I venture to say that India followed your lead to the best of her ability and this I think can hardly be denied, and still do you mean to say that the experiment was not tried far enough?

"In fact, your programme failed for no fault of the Indian. You gave only a programme to the country, but you could not lead the nation to a victorious end. To say that non-violent non-cooperation failed because the people were not sufficiently non-violent is to argue like a lawyer and not like a prophet. The people could not be more non-violent than they were during the last few years. I would like to say that they were non-violent to a degree which smelt of cowardice. You would perhaps say that it was not this non-violence, the non-violence of the cowardice that you wanted. But your programme did not contain that item which could transform cowards into heroes or which could detect and ultimately reject the cowards from the bands of heroes. This was no fault of the people. And to say that the majority of non-cooperators were cowards and not heroes is to shirk responsibilities. To say this is rather to commit an outrage on the manliness of the nation. Indians are not cowards. Their heroism can always be compared with that of the best heroes of the world. To deny this is to deny history. When I speak of India's heroism, I mean not only the heroism which sparkled in the annals of the glorious past, but I include the heroism that is manifesting itself in the present, because India is still not dead.

"What India wants is a true leader, a leader like Guru Gobind Singh or Guru Ramdas and Shivaji. India wants a Krishna who can give a worthy ideal, to be followed not by India alone, but by all humanity, by all the members of this humanity with diverse temperaments and capacities.

"Non-violent Non-cooperation Movement failed not because of there sporadic outburst of suppressed feelings here and there but because the movement was lacking in a worthy ideal. The ideal that you preached was not in keeping with Indian culture and traditions. It savoured of imitation. Your philosophy of non-violence, at least the philosophy that you gave to the people for their acceptance, was a philosophy arising out of despair. It was not the spirit of *kshama* of the Indian rishis, it was not the spirit of ahimsa of the great Indian yogins. It was an imperfect physical mixture of Tolstoyism and Buddhism and not a chemical mixture of East and West. You adopted the western method of congresses and conferences and tried to persuade the whole nation to accept the spirit of *ahimsa*, irrespective of desh, kal and patra like Tolstoy, but which was a matter of individual *sadhana* with the Indians. And above all, you were and are still vague as regards India's ultimate political goal. This is miserable. Your idea of independence is not in consistence with Indian ideals. India stands for '*sarvam paravasham dukkham sarva matmavasham sukham*' and for the ideal that individual existence is solely for the purpose of humanity and through humanity serving god. '*Jagathitaya cha krishnaya cha*'. The non-violence that India preaches is not non-violence for the sake of non-violence, but non-violence for the good of humanity, and when this good for humanity will demand violence and bloodshed, India will not hesitate to shed blood just in the same way a surgical operation necessitates the shedding of blood. To an ideal Indian, violence or nonviolence has the same significance provided they ultimately do good to humanity. '*Vinashay cha dushkritama*' was not spoken in vain.

"To my mind, therefore, the ideal that you gave to the nation or the programme of action that you laid before it is neither consistent with Indian culture nor practicable as a political programme.

"It is simply inconceivable and incomprehensible to think that you still dare to entertain the slightest hope that England can be just and generous out of her free will—this England 'which believes in Jallianwalabagh massacres as a legitimate means of self-defence', this England which tried the O'Dwyer-Nair case and gave judgement in favour of barbarism. If you have an iota of faith left in you in the good sense of the British government, then according to you where is the necessity of any programme at all? If there is any necessity of any movement in order to bring the British government to their senses, then why speak of the honesty and good intentions of the

British government? It seems that the prophet in you is gone and you are once more a lawyer defending a weak case; or perhaps you are always an exponent—a mighty exponent—of half truths only. A sovereign independent Indian Republic in alliance or in federation with the other independent nations of the earth is one thing, and self-governing India within this imperialistic British emphasis, perfectly another thing. Your sentiment of remaining within the British empire reminds one of the many Himalayan miscalculations that you have compromised a worthy ideal with the present needs of a false expediency and this is the reason that you have failed to capture the imagination of the youths of the country, youths who could dare and who are still daring to go against your wishes, although they unhesitatingly recognise you as one of the greatest of personalities of the modern age. These are the Indian revolutionaries. They have now decided to remain silent no more and therefore they request you to retire from the political field or else to direct the political movement in a way so that it may be a help and not a hindrance to the revolutionary movement. They suspended their activities so long simply to comply to your requests direct and indirect, and they went further. They actually helped you in the carrying out of your programme to the best of their abilities. But now the experiment is over and therefore the revolutionaries are free from their promise, or as a matter of fact, they promised to remain silent only for a year and no more.

"Further, I would like to point out that you have misjudged the revolutionaries in many respects when you blamed them in your recent presidential address in the 39th Congress. You said that the revolutionaries are retarding India's progress. I do not know what you mean by this word 'progress'. If you mean political progress, then can you deny that every political progress that India has already made, however little that might be, has been made chiefly by the sacrifices and the efforts of the revolutionary party? Can you deny that the Bengal partition was annulled through the efforts of the Bengal revolutionaries? Can you doubt that the Morley-Minto reform was the outcome of the Indian revolutionary movement which was mainly, though not wholly, instrumental in bringing about the Montford reform?, I shall not be very much surprised if you will answer these queries in the affirmative but I can assure you that the British government realises the potentiality of this movement. Even the late Mr. Montagu expressed to an Indian of position and rank that he took the trouble of coming to India and risked his life simply due to the activities of the young Indian revolutionaries.

"If you mean that these reforms are no index to true progress, then I would venture to say that this revolutionary movement has achieved no mean progress in the moral advancement of India. Indians were miserably afraid of death and this revolutionary party once more made the Indians realise the grandeur and the beauty that lie in dying for a noble cause. The revolutionaries have once again demonstrated that death has a certain charm and is not always a dreadful thing. To die for own beliefs and convictions, to die in the consciousness that by so dying one is serving god and the nation, to accept death or to risk one's life when there is every probability of death, for a cause which one honestly believes to be just and legitimate—is this no moral progress?

"To cling to one's cherished ideal even in adversity and temporary failures—not to be swayed away by temporary excitements and by the seemingly noble doctrines of an alluring personality, not to be daunted by long terms of imprisonment with hard labour, to be true to one's own self for years together—is this tenacity of purpose, this sturdiness in the character no index to true moral progress that India has made? And is this not the manifest outcome of the revolutionary ideal?

"You have said to the revolutionaries, 'You may not care for your own lives, but you dare not disregard those of your countrymen who have no desire to die a martyr's death'. But the revolutionaries are at a sad loss to understand the meaning of this sentence. Do you mean to say that the revolutionaries are responsible for the deaths of 70 men who were condemned in the Chauri Chaura trial? Do you mean to say that the revolutionaries are responsible for the bombing and killing of innocent people at Jallianwalabagh and Gujranwalla? Did the revolutionaries during their struggle for the last twenty years, in the past or in the present, ever ask the starving millions to take part in the revolutionary struggle? The revolutionaries have perhaps a better knowledge of the mass psychology than most of the present leaders. And this was the reason that they never wanted to deal with the masses until they became sure of their own strength. They always believed that the masses of Northern India were ready for any emergency and they were also right in thinking that the masses of Northern India as a dense matter of high explosive, dangerous to be handled carelessly. It was you and your lieutenants who misjudged the sentiments of the masses and dragged them into the *satyagraha* movement, people who were groaning under a thousand oppressions from within and without, where the lightning of anger lay unperceived and you had to pay the penalty for it. But can you give any instance where the revolutionaries dragged unwilling souls into the valley of death?

"But if you mean by the sentence that innocent people are being harassed, imprisoned and put to death due to the activities of the revolutionaries, then I would unhesitatingly and honestly admit, as far as my knowledge goes, that not a single individual was hanged who was innocent of any revolutionary activity; and about imprisonments and tortures, I may say that many innocent men were actually harassed and put to torture. But can the revolutionary party be made responsible for the atrocities committed by a foreign government? The foreign government is determined to crush any manifestation of manhood in the nation, in any form whatsoever; but in so crushing the government is very liable to commit blunders and harass and imprison and put to torture cowards along with the heroes; but are the brave people to be blamed for the sufferings of the cowards? Moreover these sufferings cannot be termed as martyrs' death.

"Lastly, I would like to say something about the remarks you have made in connection with the strength of the British empire. You have said to the revolutionaries. 'Those whom you seek to depose, are better armed and infinitely better organised than you are'. But is it not shameful that a handful of Englishmen are able to rule India, not by the free consent of the Indian people but by the force of the sword? And if the English can be well armed and well organised, why cannot the Indians be better armed and better organised? Still Indians who are saturated with the high principles of spirituality? Indians are men in the same sense as the Englishmen are. Then, what on earth makes the Indians so helpless as to think that they can never be better organised than their English masters? By what argument and logic of fact can you disprove the possibilities in which the revolutionaries have immense faith? And the spirit of non-violence that arise out of this sense of helplessness and despair can never be the non-violence of the strong, the non-violence of the Indian rishis. This is tamas, pure and simple.

"Excuse me Mahatmaji, if I am severe in criticising your philosophy and principles. You have criticised the revolutionaries most unsympathetically and even you went so far as to describe them as the enemies of the country, simply because they differ from your views and methods. You preach tolerance but you have been violently intolerant in your criticisms of the revolutionaries. The revolutionaries have risked their everything to serve their motherland, and if you cannot help them, at least be not intolerant towards them."[3]

I never made any promise to anybody as to when and how I should retire from the political life of the country. But I did say and now repeat that I would certainly retire if I find that India does not imbibe my message and that India wants a bloody revolution. I should have no part in that movement because I do not believe in its utility either for India, or, which is the same thing, for the world.

I do believe that there was a wonderful response to the call of non-cooperation, but I do also believe the success was more than proportionate to the measure of non-cooperation. The wonderful awakening of the masses is a standing demonstration of the fact. I do believe too, that the country exercised great self-restraint; but I must reiterate my opinion that the observance of non-violence was far below the required standard.

I do not believe that "my philosophy" is an indifferent mixture of Tolstoy and Buddha. I do not know what it is except that it is what I feel to be true. It sustains me. I owe much to Tolstoy and much to Buddha. I still somehow or other fancy that "my philosophy" represents the true meaning of the teachings of the *Gita*. I may be totally mistaken. Such a mistake can do no harm either to me or to anybody. For the source of my inspiration is of no consequence if what I stand for be unadulterated truth.

Let the philosophy I represent be tested on its own merits. I hold that the world is sick of armed rebellions. I hold too that whatever may be true of other countries, a bloody revolution will not succeed in India. The masses will not respond. A movement in which masses have no active part can do no good to them. A successful bloody revolution can only mean further misery for the masses. For it would be still foreign rule for them. The non-violence I teach is active non-violence of the strongest. But the weakest can partake in it without becoming weaker. They can only be stronger for having been in it. The masses are far bolder today than they ever were. A non-violent struggle necessarily involves construction on a mass scale. It cannot therefore lead to tamas or darkness or inertia. It means a quickening of the national life. That movement is still going on silently, almost imperceptibly, but none the less surely.

I do not deny the revolutionary's heroism and sacrifice. But heroism and sacrifice in a bad cause are so much waste of splendid energy and hurt the good cause by drawing away attention from it by the glamour of the misused heroism and sacrifice in a bad cause.

I am not ashamed to stand erect before the heroic and self-sacrificing revolutionary, because I am able to pit an equal measure of non-violent men's heroism and sacrifice untarnished by the blood of the innocent. Self-sacrifice of one innocent man is a million times more potent than the sacrifice of million men who die in the act of killing others. The willing sacrifice of the innocent is the most powerful retort to insolent tyranny that has yet been conceived by god or man.

I invite the attention of the revolutionaries to the three great hindrances to swaraj—the incomplete spread of the spinning wheel, the discord between Hindus and Musalmans and the inhuman ban upon the suppressed classes. I ask them patiently to take their due share in this work of patient construction. It may not be spectacular enough. But on that very account it requires all the heroic patience, silent and sustained effort, and self-effacement of which the tallest among the revolutionaries is capable. Impatience will blur the revolutionary's vision and lead him astray. Slow and inglorious self-imposed starvation among the starving masses is every time more heroic than the death on the scaffold under false exaltation.

All criticism is not intolerance. I have criticised the revolutionary because I have felt for him. He has the same right to hold me to be in error as I believe him to be in error.

There are other points that are covered by the "open letter". But I have omitted to refer to them because I think that they can be easily answered by the reader and in no case do they touch the vital issue.[4]

As Sachindranath Sanyal was arrested about this time, he was not there to carry on the controversy. I waited for some time, then I took up the gauntlet. I wrote a letter and signed it the same way "A Revolutionary".[5] It was published in *Young India* on 9 April 1925 with Gandhi's reply.

"The Revolutionary whom I endeavoured to answer some time ago, has returned to the charge and challenges me to answer certain questions that arise out of my previous answers to him. I gladly do so. He seems to me to be seeking light even as I am and argues fairly and without much passion. So long as he continues to reason calmly I promise to continue the discussion. His first question is:

"Do you really believe that the revolutionaries of India are less sacrificing, less noble or less lovers of their country than the swarajists, moderates and the nationalists? May I challenge you to keep before the public the names of some swarajists, moderates or nationalists who have embraced the death of a martyr for the sake of the motherland? Can you be bold, nay, arrogant enough to deny it in the face of historical facts that the revolutionaries have sacrificed more for their country than any other party which professes to serve India? You are ready to make compromises with other parties, while you abhor our party and describe the sentiments as poison. Will you not tremble to use the same word of intolerance for the sentiments of any other party which is decidedly inferior in the eyes of god and man to us? What makes you shrink from calling them misguided patriots or venomous reptiles?"

"I do not regard the revolutionaries of India to be less sacrificing, less noble or less lovers of their country than the rest. But I respectfully contend that their sacrifice, nobility and love are not only a waste of effort, but being ignorant and misguided, have done more harm to the country than any other activity. For the revolutionaries have retarded the progress of the country. Their reckless disregard of the lives of their opponents has brought on repression, that has made those that do not take part in their warfare more cowardly than they were before. Repression does good only to those who are prepared for it. The masses are not prepared for the repression that follows in the trial of revolutionary activities and unwittingly strengthen the hands of the very government which the revolutionaries are seeking to destroy. It is my certain conviction that had the Chauri Chaura murders not taken place, the movement attempted at Bardoli would have resulted in the establishment of swaraj. Is it, therefore, any wonder that with such opinion I call the revolutionary a misguided and therefore dangerous patriot? I would call my son a misguided and dangerous nurse who because of his ignorance and blind love fought at the cost of his own physicians whose system of medicine no doubt did not but which I could not escape for want of will or ability result would be that I would lose a noble son and bring down upon my head the wrath of the physicians who suspecting my complicity in the son's activities might seek to punish me in addition to continuing their harmful course of treatment. If the son had attempted to convince the physicians of the error or me of my weakness in submitting to the treatment, the physicians might have mended their way or I might have rejected the treatment or would at least have escaped the wrath of the physicians. I do make certain compromises with the other parties because, though I disagree with them, I do not regard their activities as positively harmful and dangerous as I regard the revolutionaries. I have never called the revolutionaries "venomous reptiles". But I must refuse to fall into hysterics over their sacrifices, however great they may be, even as I must refuse to give praise to the sacrifice of my misguided son for his sacrifice in the illustration supposed by me. I feel sure that those who through sufficient reasoning or false sentiment secretly or openly give praise to the revolutionaries for their sacrifices do harm to them and the cause they have at heart. The writer has asked me to quote instances of non-revolutionary patriots who gave their lives for the country. Well, two complete cases occur to me as I write these notes. Gokhale and Tilak died for their country. They worked in almost total disregard of their health and died much earlier than they need have. There is no necessary charm about death on the gallows; often such death is easier than a life of drudgery and toil in malarious tracts. I am quite satisfied that among the swarajists and others there are men who will any day lay down their lives if they felt convinced that their death would bring deliverance to the country. I suggest to my friend, the revolutionary that death on the gallows serves the country only when the victim is a "spotless lamb".

"'India's path is not Europe's'. Do you really believe it? Do you mean to say that warfare and organisation of army was not in existence in India, before she came in contact with Europe? Warfare for fair cause—Is it against the spirit of India? *Vinashaya cha dushkritam*!—Is it something imported from Europe? Granted that it is, will you be fanatic enough not to take from Europe what is good? Do you believe that nothing good is possible in Europe? If conspiracy, bloodshed and sacrifice for fair cause are bad for India, will they not be bad as well for Europe?"

"I do not deny that India had armies, warfare etc., before she came in contact with Europe. But I do say that it never was the normal course of Indian life. The masses unlike those of Europe were untouched by the warlike spirit. I have already said in these pages that I ascribe to the *Gita*, from which the writer has quoted the celebrated verse, a totally different meaning from that ordinarily given. I do not regard it as a description of or an exhortation to physical warfare. And in any case according to the verse quoted, it is god, the all knowing, who descends to the earth to punish the wicked. I must be pardoned if I refuse to regard every revolutionary as an all-knowing god or an avatar. I do not condemn everything European. But I condemn for all climes and for all times secret murders and unfair methods even for a fair cause.

"'India is not Calcutta and Bombay.' May I most respectfully put it before your Mahatmaship that the revolutionaries know this geography of India enough to be able to know this geographical fact easily. We hold this fact as much as we hold that a few spinners do not form the Indian nation. We are entering villages and have been successful everywhere. Can you not believe that they, the sons of Shivaji, Pratap and Ranjit, can appreciate our sentiments with more readiness and depth than anything else? Don't you think that armed and conspired resistance against something satanic and ignoble is infinitely more befitting for any nation, especially India, than the prevalence of effortlessness and philosophical cowardice? I mean the cowardice which is pervading the length and breadth of India, owing to the preaching of your theory of non-violence or more correctly the wrong interpretation and misuse of it. Non-violence is not the theory of the weak and helpless, it is the theory of the strong. We want to produce such men in India, who will not shrink from death whenever it may come and in whatever form—will do the good and die. This is the spirit with which we are entering the villages. We are not entering the villages to extort votes for councils and district boards, but our object is to secure co-martyrs for the country who will die and stone will not tell where his poor corpse lies. Do you believe like Mazzini that ideas ripen quickly, when nourished by the blood of martyrs?"

It is not enough to know the geographical difference between Calcutta and the villages outside the railways. If the revolutionaries knew the organic difference between these, they would, like me, become spinners. I own that the few spinners we have not make India. But I claim that it is possible to make all spin as it did before, and so far as sympathy is concerned millions are even now in sympathy with the movement, but they never will be with the revolutionary. I dispute the claim that the revolutionaries are succeeding with the villagers. But if they are, I am sorry. I shall spare no pains to frustrate their effort. Armed conspiracies against something satanic is like matching satans against satan. But since one satan is one too many for me, I would not multiply him. Whether my activity is effortlessness or all efforts remain perhaps to be seen. Meanwhile, if it has resulted in making two yards of yarn spin where only one was spinning, it is so much to the good. Cowardice, whether philosophical or otherwise, I abhor. And if I could be persuaded that revolutionary activity has dispelled cowardice, it will go a long way to soften my abhorrence of the method, however much I may still oppose it on principle. But he who runs may see that owing to the non-violent movement, the villagers have assumed a boldness to which only a few years ago they were strangers. I admit that non-violence is a weapon essentially of the strong. I also admit that often cowardice is mistaken for non-violence.

My friend begs the question when he says a revolutionary is one who "does the good and dies". That is precisely what I question. In my opinion he does the evil, and dies. I do not regard killing or assassination or terrorism as good in any circumstances whatsoever. I do believe that ideas ripen quickly when nourished by the blood of martyrs. But a man who dies slowly of jungle fever in service bleeds as certainly as the one on the gallows. And if the one who dies on the gallows is not innocent of another's blood, he never had ideas that deserved to ripen.

"One of your objections against the revolutionaries is that their movement is not mass movement, consequently the mass at large will be very little benefited by the revolution, for which we are preparing. That is indirectly saying that, we shall be most benefitted by it. Is it really what you mean to say? Do you believe that those persons who are ever ready to die for their country, those mad lovers of their country, I mean the revolutionaries of India in whom the spirit of *nishkama karma* reigns, will betray their motherland and secure privileges for a life—this trifling life? It is true that we will not drag the mass just now in the field of action, because we know that it is weak, but when the preparation is complete we shall call them in the open field. We profess to understand the present Indian psychology full well, because we daily get the chance of weighing our brethren along with ourselves. We know that the mass of India is after all Indian, it is not weak by itself but there is want of efficient leaders; so when we have begot the number of leaders by constant propaganda and preaching, and the arms, not shrink from calling, and if necessary, dragging the mass in the open field to prove that they are the descendants of Shivaji, Ranjit, Pratap and Gobind Singh. Besides, we have been constantly preaching that the masses are not for the revolution but the revolution is for the masses. Is it sufficient to your prejudice in this connection?"

I neither say nor imply that the revolutionary benefits if the masses do not. On the contrary, and as a rule, the revolutionary never benefits in the ordinary sense of the word. If the revolutionaries succeed in attracting, not "dragging" the masses to them, they will find that the murderous campaign is totally unnecessary. It sounds very pleasant and exciting to talk of "the descendants of Shivaji, Ranjit, Pratap and Gobind Singh". But is it true? Are we all descendants of these heroes in the sense in which the writer understands it? We are their countrymen, but their descendants are the military classes. We may in future be able to obliterate caste, but today it persists and therefore the claim put up by the writer cannot in my opinion be sustained.

"Last of all, I shall ask you to answer these questions: Was Guru Gobind Singh a misguided patriot because he believed in warfare for noble cause? What will you like to say about Washington, Garibaldi and Lenin? What do you think of Kamal Pasha and De Valera? Would you like to call Shivaji and Pratap, well meaning and sacrificing physicians who prescribed arsenic when they should have given fresh grape juice? Will you like to call Krishna Europeanised because he believed also in the *vinasha of dushkritas*?"

This is a hard or rather awkward question. But I dare not shirk it. In the first instance Guru Gobind Singh and the others whose names are mentioned did not believe in secret murder. In the second, these patriots knew their work and their men, whereas the modern Indian revolutionary does not know his work. He has not the men, he has not the atmosphere, that the patriots mentioned had. Though my views are derived from my theory of life, I have not put them before the nation on that ground. I have based my opposition to the revolutionaries on the sole ground of expedience. Therefore, to compare their activities with those of Guru Gobind Singh or Washington or Garibaldi or Lenin would be most misleading and dangerous. But by test of the theory of non-violence I do not hesitate to say that it is highly likely that had I lived and in the respective countries I would as their contemporary have called every one of them a misguided Patriot, even though a successful and brave warrior. As it is, I must not judge them. I disbelieve history so far as details of acts of heroes are concerned. I accept broad facts of history and draw my own lessons for my conduct. I do not want to repeat it in so far as the broad facts contradict the highest laws of life. But I positively refuse to judge men from the scanty material furnished to us by history. *De mortuis nil nisi bonum*. Kamal Pasha and De Valera, too, I cannot judge. But for me as a believer in non-violence out and out, they cannot be my guides in life in so far as their faith in wår is concerned. I believe in Krishna perhaps more than the writer. But my Krishna is the lord of the universe, the creator, preserver and destroyer of us all. He may destroy because he creates. But I must not be drawn into a philosophical or religious argument with my friends. I have not the qualifications for teaching my philosophy of life. I have barely qualifications for practising the philosophy I believe. I am poor but a struggling soul yearning to be wholly good, wholly truthful and wholly non-violent in thought, word and deed, but even failing to reach the ideal which I know to be true. I admit and assure my revolutionary friends, it is a painful climb but the pain of it is a positive pleasure for me. Each step upward makes me feel stronger and fit for the next. But all that pain and the pleasure are for me. The revolutionaries are at liberty to reject the whole of my philosophy. To them I merely present my own experience as a co-worker in the same cause even as I have successfully presented them to the Ali Brothers and many other friends. They can and do applaud whole heartedly the action of Mustafa, Kamal Pasha and possibly De Valera and Lenin. But they realise with me that India is not like Turkey or Ireland or Russia and that revolutionary activity is suicidal at this stage of the country's life at any rate, if not for all time in a country so vast, so hopelessly divided and with the masses so deeply sunk in pauperism and so fearfully terror struck.[6]

Sanyal wrote another letter. This was also published in *Young India* in the same way on 7 May 1925. Here is the letter as published in the *Young India*:

"My revolutionary friend has returned to the charge, but I must tell him that he has not been as patient with his composition as before. He has introduced in his letter under discussion much irrelevant matter and has argued loosely. So far as I can see, he has exhausted all his argument and has nothing new to say. But should he write again, I advise him to write his letter more carefully and boil down his thoughts. I have been obliged to do that for him this time. But as he is see king light, let him read carefully what I write, then think out his thoughts calmly and write out clearly and briefly. If it is merely questions he has to ask let him simply write them out without arguing to convince me. I do not pretend to know everything about the revolutionary movement, but as I have been obliged to think, observe and write a great deal, there is very little new that he can tell me. Whilst, therefore, I promise to keep an open mind, I ask him, please, to spare a busy servant of the nation and a true friend of the revolutionary the labour of reading much that he need not read. I am anxious to keep in touch with the revolutionary and I can only do so through these columns. I have a soft corner for him in my heart for there is one thing in common between him and me—the ability to suffer. But as I humbly believe him to be mistaken and misguided, I desire to wean him from his error or in the process myself be weaned from mine.

My revolutionary friend's first question is:

"'The revolutionaries have retarded the progress of the country. Do you differ with your own view, when you wrote in connection with the Bengal partitions: 'After the partition people saw that petitions must be backed up by force, and that they must be capable of suffering. This spirit must be considered to be the chief result of the partition... That which the people said tremblingly and in secret began to be said and written openly... people, young and old, used to run away at the sight of an English face; it no longer awed them. They did not fear even a row, or being imprisoned. "Some of the best sons" of India are at present in banishment. The movement which followed the partition or more correctly which was the manifestation of the unrest of the people was the revolutionary movement, and the best sons of India you speak of are mostly revolutionaries or semi-revolutionaries. How is it that these so-called ignorant and misguided persons were able to reduce it, not remove the cowardice of India? Would you be so intolerant as to call the revolutionaries ignorant, because they cannot understand your peculiar dogma of non-violence?"

There is no difference between the view expressed in Indian Home Rule from which the writer has quoted and the views now expressed by me. Those who led the partition movement, whatever and whoever they were, undoubtedly shed the fear of Englishmen.

That was a distinct service to the country. But bravery and self-sacrifice need not kill. Let my friend remember that Indian Home Rule as the booklet itself states was written in answer to the revolutionary's arguments and methods. It was an attempt to offer the revolutionary something infinitely superior to what he had, retaining the whole of the spirit of self-sacrifice and bravery that was to be found in the revolutionary. I do not call the revolutionary ignorant, merely because he does not understand or appreciate my method, but because he does not even appear to me to understand the art of warfare. Everyone of the warriors whom my friend quotes knew his art and had his men.

The second question is:

"Was Terence MacSwiney a 'spotless lamb' when he died of hunger strike of 71 days? Please remember that he was to the last an advocate of conspiracy, bloodshed and terrorism, and maintained his ideas expressed in his famous book *Principles of Freedom*. If you can call MacSwiney a 'spotless lamb', will you not be ready to use the same term for Gopimohan Saha?"

I am sorry to say I do not know enough of the life of MacSwiney to be able to give an opinion. But if he advocated "conspiracy, bloodshed and terrorism", his method was open to the same objections that have been advanced in these pages. I never regarded him as a "spotless lamb". I gave my humble opinion when his fast was declared, that from my standpoint it was an error. I do not justify every fast.

The third question is:

"You believe in *varnas*. Therefore, it is self-evident that you hold the *kshatriyas* to be of the same utility as any other varna. The revolutionaries profess to be kshatriyas in this *nihkshatriya* epoch in India: '*Kshatat trayate iti kshatriya*'. I consider this state of India to be the greatest *kshata* which India has ever met with. In other words this is the time when the need of kshatriyas in India is the uttermost. Manu, the prince of Hindu Law givers, prescribes four ways for the kshatriya: *'sama, dama, danda, bheda'*. In this connection I reproduce a passage from Vivekananda which I think will greatly help you to comprehend the matter fully well.

"All great teachers have taught "resist not evil", they have taught that non-resisting is the highest moral ideal. We all know that if, in the present state of world, people try to carry out this doctrine, the whole social fabric would fall to pieces, society would be destroyed, the violent and the wicked will take possession of our property, and possibly take our lives also. Even one day of such non-resistance would lead to the utter dissolution of the country. I know what you will do in this awkward position, you will try to interpret it differently, but you shall find that he has left no room for such misinterpretation, because he instantly adds, some of you have read perhaps the *Bhagvad Gita* and many of you in western countries may have felt astonished at the first chapter wherein our Shri Krishna calls Arjuna a hypocrite and coward, on account of his refusal to fight or offer resistance, because his adversaries were his friends and relatives—his refusal on the plea that non-resistance was the highest ideal of love. There is a great lesson for us all to learn, that in all things the two extremes are alike; the extreme positive and the extreme negative are always similar; when the vibrations of light are too slow we do not see them nor do we see them when they are too rapid; so also with sound, when very-low in pitch we do not hear it, when very high we do not hear it either, of like notion is the difference between resistance and non-resistance.... We must first care to understand whether we have the power of resistance or not. Then having the power if we renounce it and do not resist we are doing a grand act of love; but if we cannot resist and yet at the same time make it appear and ourselves believe that we are actuated by motives of highest love, we shall be doing the exact opposite of what is morally good. Arjuna, became coward at the sight of the mighty array against him, his "love" made him forget his duty towards his country and king. That is why Shri Krishna told him that he was a hypocrite: "Thou talkest like a wise man, but thy actions betray thee to be a coward, therefore stand up and fight"."

I want to add nothing more except a few questions. Do you think that your so-called heart and soul non-violent disciples can resist this alien bureaucratic government by physical force? If yes, on what ground; if not how then does your non-violence remain the weapon of the strong? Please answer these questions in the most unmistakable terms, so that no one can make different interpretations.

"Along with it I shall ask you the following questions, which directly arise from your statement. In your swarajya, is there any place for soldiers? Will your swarajya government keep armies? If so, will they fight, I mean use physical force, when necessary, or will they offer *satyagraha* against their opponents?"

I have room in my philosophy of life for kshatriyas. But my definition of him I take from the *Gita*. He who does not run away from battle, i.e. danger is a kshatriya. As the world progresses, the same terms acquire new values. Manu and the other lawgivers did not lay down eternal principles of conduct. They enunciated certain eternal maxims of life and laid down for their age rules of conduct, more or less in accord with those maxims. I am unable to subscribe to the methods of bribery and deceit, even for gaining entrance into heaven, much less for gaining India's freedom. For heaven will not be heaven and freedom will not be freedom if either is gained through such methods.

I have not verified the quotation said to be from Vivekananda. It has neither the freshness nor the brevity that mark most of that great man's writings. But whether it is from his writings or not, it does not satisfy me. If a large number of people carry out the doctrine of non-resistance, the present state of the world will not be what it is. Those individuals who have carried it out have not lost anything. They have not been butchered by the violent and the wicked. On the contrary the latter have shed both their violence and wickedness in the presence of the nonviolent and the good.

I have already stated my meaning of the *Gita*. It deals with the eternal dual between good and evil. And who does not, like Arjuna, often quail when the dividing line between evil is thin and when the right choice is so difficult.

I heartily endorse, however, the statement that he alone is truly non-violent who remains non-violent even though he has the ability to strike. I do therefore claim that my disciple (I have only one and that is myself) is quite capable of striking, very indifferently and perhaps ineffectively I admit; but he has no desire to do so. I have had in my life many an opportunity of shooting my opponents and earning the crown of martyrdom but I had not the heart to shoot any of them. For I did not want them to shoot me, however, much they disliked my methods. I wanted them to convince me of my error as I was trying to convince them of theirs. "Do unto others as you would that they should do unto you."

Alas! in my swaraj of today there is room for soldiers. Let the revolutionary friend know that I have described the disarmament and consequent emasculation of a whole people as the blackest crime of the British. I have not the capacity for preaching universal non-violence to the country. I preach therefore non-violence restricted strictly to the purpose of winning our freedom and therefore perhaps for preaching the regulation of international relations by non-violent means. But my incapacity must not be mistaken for that of the doctrine of non-violence. I see it with my intellect in all its effulgence. My heart grasps it. But I have not yet the attainments for preaching universal non-violence with effect. I am not advanced enough for the great task. I have yet anger within me, I have yet the *dwait bhava,* duality in me. I can regulate my passions, I keep them under subjection, but before I can preach universal nonviolence with effect, I must be wholly free from passions. I must be wholly incapable of sin. Let the revolutionary pray with and for me that I may soon become that. But meanwhile let him take with me the one step to it which I see as, clearly as day light, i.e. to win India's freedom with strictly non-violent means. And then under swaraj, you and I shall have a disciplined, intelligent, educated police force that would keep order within and fight raiders from without if by that time I or someone else does not show a better way of dealing with either.[7]

These letters created a great commotion at that time as they were quoted in all important newspapers. Now there were definitely two camps inside the national movement. The revolutionaries from the very beginning were ready to tolerate Gandhi and even consider his activities as supplementing the other activities, but Gandhi and to some extent Jawaharlal Nehru, although admiring and praising De Valera of Ireland and Kamal Pasha of Turkey, wanted to wipe away the revolutionaries.

Apart from the above documents a glimpse of the ideology of revolutionaries could be had from a model list of books or revolutionaries prepared by Sanyal. This list contained books by or on De Valera, Garibaldi, Mazzini. There were some books on Russia wrote Sanyal, "I do not remember the names. I bought a certain book *Russ Ki Rajyakranti* (Russian Revolution) in Hindi and many other books. At that time I read one book *History of Socialism* by Kirkup. I think during that period that was the only book on the history of socialism available in India. Then there were books in Bengali, English and Hindi on various patriotic subjects."[8]

Ideas of scientific socialism were fast invading the minds of Indian youths. But while reading the history of the Russian Revolution we did not distinguish between the activities of the Bolsheviks and the Narodniks. To us, Lenin was a good patriot like Garibaldi or Sun Yat Sen. I had heard and read about Marx, but failed to understand at that time his role fully. He more or less seemed to be an extremely kind hearted gentleman, philanthropically inclined towards the proletariat. His beard and his eyes inspired us with great respect, but this respect was not very much different from that we entertained towards say, Rabindranath Tagore.[9]

This was not very strange seeing that for us Indians in 1922-25, the real fight was against the British. We had to win the patriotic war, of course along with it we could fight or at least prepare for the fight for socialism. That is a different question.

REFERENCES

1. Gupta, Manmath Nath, *They Lived Dangerously: Reminiscences of a Revolutionary* (Delhi: Peoples Publishing House), 1969, p. 69.
2. *Young India,* 12-02-1925.
3. *Ibid.*
4. *Ibid.*
5. Gupta, Manmath Nath, *They Lived Dangerously: Reminiscences of a Revolutionary, op. cit.*, p. 78.
6. *Young India*, 09-04-1925.
7. *Ibid.*, 07-05-1925.
8. Gupta, Manmath Nath, *They Lived Dangerously: Reminiscences of a Revolutionary, op. cit.*, p. 89.
9. *Ibid.*

6

COULD GANDHI HAVE SAVED BHAGAT SINGH?

The bold act of throwing bombs in the Central Assembly and the arrests of Bhagat Singh and Batukeshwar Dutt had attracted people's attention towards them. Instead of escaping from the spot, which was not a difficult task for both of them, they stood there like a rock, raised revolutionary slogans, threw leaflets and willingly courted arrest. This act made Bhagat Singh a symbol of the nation, a symbol to be honoured and responded and also to be emulated by the youths of India.[1] According to Jawaharlal Nehru, within a few months each town and village of Punjab and to a lesser extent the rest of northern India resounded with his name. Innumerable songs grew up about him and the popularity that the man achieved was something amazing.[2] The manner in which Bhagat Singh and Dutt behaved during their trial on the Delhi Assembly Case also endeared Bhagat Singh and his comrade to the nation. 'Inquilab Zindabad', the slogan coined by Bhagat Singh and raised throughout the proceedings of the Delhi Assembly Bomb Case by him and Dutt became so much popular that its echo was heard in each and every part of the country.

On witnessing the nation's admiration for Bhagat Singh the British government was alarmed. It was conscious of the fact that its image had been tarnished by the one-sided trial in the Assembly Case, firstly before the magistrate and later before the Session Judge. Bhagat Singh became the main target of the government but he had no fears. In order to improve the plight of the political prisoners in the Indian jails and to draw the attention of the people, he and Dutt commenced their hunger-strike in Mianwali and Lahore jails respectively. After his transfer to Lahore Jail, Bhagat Singh continued the hunger strike. They were joined by other co-accused of the Lahore conspiracy case confined in Central and Borstal Jails of Lahore. Although the hunger-strike was suspended and resumed several times, Bhagat Singh and his comrades were able to win the sympathy and support of the people for this noble cause. In the Legislative Assembly Pandit Moti Lal Nehru criticised the government's policy towards the under trials in the Lahore conspiracy case. Jawaharlal Nehru also expressed his concern and visited the two jails of Lahore.

During the trial of Lahore conspiracy case, Bhagat Singh disseminated his views frankly and fearlessly. Besides explaining the philosophy of the revolutionaries he drew the attention of the public towards the unjust actions of the government. He was successful to a great extent in achieving both the objectives. Crowd of people came to watch the proceedings in the Court. At the main gate of the court, facing towards the road, a large number of students of schools and colleges always gathered to witness the proceedings. The huge crowd of the youths were often heard singing patriotic songs like '*Kabhi woh din bhi ayega ke jab azad hum honge*', '*Watan par marne walon ka baki nishan hoga*' and '*Sarfaroshi ki tamanna ab hamare dil mein hai*'.[3] Such popular outbursts were also witnessed in almost all cities of the country. Undoubtedly at that time Bhagat Singh was the most popular figure of the nation.

In this tense atmosphere the verdict of death passed by the Tribunal on Bhagat Singh, Sukhdev and Rajguru virtually shook the nation. In a speech at Allahabad on October 12, 1930, five days after the award of sentence of death, Jawaharlal Nehru said, "Whether he agreed with Bhagat Singh or not, his heart was full of admiration for the courage and self-sacrifice of Bhagat Singh. A man of his type is exceedingly rare. If the Viceroy expects us to refrain from admiring this wonderful courage and high purpose behind it, he is mistaken. Let him ask his own heart what he would have felt if Bhagat Singh had been an Englishman and acted for England."[4]

Meanwhile in an application, made on behalf of most of the accused, permission to appeal to the Privy Council was sought.[5] Bhagat Singh did not associate himself with this application. In fact during the period between the order of death sentence on October 7, 1930 and his execution on March 23, 1931, he always opposed the idea of submission of an appeal for mercy to the Privy Council.

The Government felt that on account of the submission of an application on behalf of the convicted revolutionaries, the execution of their sentences was likely to be delayed for some time. In this intervening period the Government of India apprehended popular demonstrations against the government. Keeping this in view, the Home Secretary, Government of India, informed all the local governments on November 4, 1930 that the Punjab Government will communicate the date of the execution of the sentences to other local governments in case the proceedings before the Privy Council were unfavourable to the accused.[6]

In view of the Punjab Government's reluctance to prohibit meetings, since such type of action would be difficult to enforce without a series of clashes with the police, on December 13, 1930 the Home Department of Government of India advised all the local governments and administrations to follow a uniform policy, keeping in view the local conditions and practices. The local governments were authorised to prohibit political meetings in sympathy with the convicts in case such a policy was being followed by them.[7]

Till December 23, 1930 there was no violent incident connected with the protests against the sentences given to the accused in the Lahore conspiracy case. On that day two students fired six shots at the Governor of Punjab who had come to address the Convocation of the Punjab University at Lahore. One of the assailants committed suicide just after firing the shots but the other one was arrested by the police. In this incident the Governor received minor injuries.[8]

In the meantime, the political developments in the country were moving at a fast speed. After the First Round Table Conference held in Lahore in November, 1930 the British Government was in a fix. The Congress had boycotted it and other parties had made separate demands for their representation. On January 26, 1931

Mahatma Gandhi was released from the jail. In the beginning Gandhiji was reluctant to talk with the Viceroy but on February 14, 1931 he decided to have a talk with Lord Irwin, the Viceroy of India for a compromise. The talks commenced on February 17, 1931.

Five days before the talks held between Gandhi and Irwin, the Secretary of State for India telegraphically informed the Viceroy that the appeal on behalf of the convicts in the Lahore conspiracy case was heard by the Judicial Committee of the Privy Council. The Committee had decided to recommend the dismissal of the appeal to His Majesty the King.[9] In view of the prospects of a compromise with Gandhiji acting on behalf of the Congress, the British Government decided to defer the execution of Bhagat Singh and his two associates.

On February 14, 1931 Madan Mohan Malviya sent a telegram to the Viceroy. He appealed to the Viceroy to exercise his prerogative of mercy in the case of Bhagat Singh, Rajguru and Sukhdev and to commute the sentences of death passed upon them into that of transportation of life. According to him, their lives should be spared not only on grounds of humanity but also in view of the fact that their action was not prompted by any personal or selfish consideration but by a patriotic impulse. Moreover it would be a great shock to the public opinion in the country. In his opinion such an act would produce a very beneficial effect on it.[10]

The leaders and the public continued to appeal to the Government to commute the sentence of death passed by the Tribunal on Bhagat Singh, Rajguru and Sukhdev. On February 18, 1931 the public submitted memorial to the Viceroy. The main arguments given in the memorial were : (i) the trial was not held according to the ordinary form of law before an ordinary court of Law, but was conducted *ex parte* in the absence of the accused by an extraordinary court of Law; (ii) after the judgement the accused were not permitted to appeal to the High Court; and (iii) Sardar Kishan Singh was not allowed to produce documentary and oral evidence to disprove the story regarding the murder of Saunders. Hence it appealed for the commutation of the death sentence.[11]

Much against the wishes of Bhagat Singh, his mother Vidyavati sent a telegram on February 17, 1931 to the Viceroy requesting the stay of execution of her son.[12] After two days she submitted a petition to the Viceroy. In this petition she did not raise any legal or political argument but put forward the following pleas : (i) in case proper cross-examination of her son had been conducted, his innocence would have been established, and (ii) under the form of trial prescribed by the Ordinance, his son was deprived of the right of appeal to the High Court, which he would have had under the Criminal Procedure Code. In the end she requested the Viceroy to exercise his prerogative of mercy for commuting the death sentence and sentence him to a term of imprisonment.[13]

On February 17, 1931 Gandhiji in his talks with the Viceroy discussed the matter about the suspension of the sentence of execution of Bhagat Singh. According to Gandhiji, he told the Viceroy, "This has no connection with our discussion and it may even be inappropriate on my part to mention it. But if you want to make the present atmosphere more favourable, you should suspend Bhagat Singh's execution." "The Viceroy expressed his gratefulness to Gandhiji for putting this matter before him in such a manner. He said, "Commutation of sentence was a difficult thing but suspension could be considered." While praising Bhagat Singh's bravery, Gandhiji said that in his opinion he was not in a right frame of mind. He then referred to the evil of capital punishment which did not give any opportunity to such a man to reform himself. He was, therefore, putting this matter on humanitarian grounds and desired suspension of the sentence in order to avoid unnecessary turmoil in the country.[14]

The disappointment of the public was shared by Subhas Bose, and a few other leaders of the Congress. A few young enthusiasts of the Congress distributed pamphlets in which they expressed their views. They asked Gandhiji as to how there can be peace when a sentence of death is hanging over the heads of these patriots? On March 7, 1931, Gandhiji addressed a mass meeting attended by over fifty thousand persons. He pointed out that two days before he had signed a provisional temporary settlement with the Viceroy which in no way can be called a peace treaty. He beseeched the young men to have courage, patience and reason. In case the old men had bungled and were guilty of weakness, the youth should force them to abdicate and assume the reins themselves. Gandhiji then explained that throughout his negotiations with the Viceroy, he was not acting on his own but was backed by the whole Working Committee of the Congress. As a negotiator of the provisional truce, the pledge of truth, non-violence and boundaries of justice were not forgotten by him. He requested the youths to stand by the settlement and secure the release of the prisoners. Along with it he issued a word of warning to the youths that Bhagat Singh cannot be released by violent means. In his place thousands of Bhagat Singhs would have to be sacrificed. As he was not prepared to do so, he preferred the way of peace and non-violence. In the end he beseeched the youths to change their methods and accept the settlement.[15]

In a speech at Allahabad on March 10, 1931 Jawaharlal Nehru threw light on the efforts made by the Congress for the release of Bhagat Singh and others. He said that in asking for the release of Bhagat Singh if the Congress had demonstrated obstinacy he would perhaps have been hanged by this time. If Bhagat Singh is alive till then it was because of Mahatma Gandhi's efforts and if he and other prisoners guilty of violence were eventually released, it would be due to Mahatma Gandhi's efforts.[16]

Although Nehru expressed a hope about the commutation of the sentence of death passed on Bhagat Singh, the Government of India was determined to carry out the decision of the Tribunal to hang Bhagat Singh and his two comrades till death.

On hearing the news about Gandhiji being in touch with the Viceroy to suspend the hanging for sometime Bhagat Singh felt some what worried. He was not inclined to pass his days any more in the dingy cell of the jail. The earlier he embraced the death, it was better for him. Moreover he had no faith in Gandhi, his cult of non-violence and the Congress. In his conversations with Chuhar Singh, a 'Ghadrite' prisoner, Bhagat Singh told him, "Babaji, the imperialist British do not care for the popular urge. The Gandhi-Viceroy meetings are on. These meetings would be no good to us. We, three will not be let free and will be hanged. The Congress policy is weak, reformatory and constitutional. The constitutionalists are always useless for the revolutionary interests. They strengthen the hands of the enemy. The opportunist enemy has made an agreement to save his skin after some give and take. When the question of implementation arises, it throws the paper on which the agreement is written into the waste paper basket. So many pacts were made previously but nothing had materialised."[17]

Gandhiji again met the Viceroy on March 19, 1931. According to the Viceroy, at the time of his departure Gandhiji asked him if he could mention the case of Bhagat Singh. Gandhiji said that he had learnt from the news in the press about Bhagat Singh's execution on March 24, 1931. According to Gandhiji it was an unfortunate day as it coincided with the arrival of the new president of the Congress at Karachi and there would be much popular excitement.

The Viceroy informed Gandhiji about his considering the case of Bhagat Singh with most anxious care but he was not able to find any grounds on which he could justify to his conscience the commutation of the sentence. As to the date he had considered the possibility of postponement till after the Congress session but had deliberately rejected the idea on the following grounds :

(i) the postponement of execution, merely on political grounds, when orders have been passed, seemed to him improper;
(ii) the postponement was inhuman in that it would suggest to Bhagat Singh's friends and relatives that he was considering commutation; and
(iii) the Congress would consider it legitimate in complaining that they were tricked by the government.

According to the Viceroy, Gandhi appeared to appreciate the force of the above arguments and did not say anything.[18]

"Did Gandhi fail to save Sardar Bhagat Singh and his comrades from the gallows?", is a question which is repeatedly asked from the historians and scholars of Freedom Movement. It is also held in certain quarters that Gandhi did not sincerely try to save them. The purpose of the present piece is not to hold any brief for the Mahatma or his approach but to put the facts straight.

Gandhi-Irwin Pact raised Gandhi's stature in the country as, for the first time the agents of British Government condescended to negotiate with the accredited agents of the Indian National Congress on equal footing for the settlement of the terms of Peace. But the execution of Bhagat Singh, Sukhdev and Rajguru neutralised this tendency. This tragedy also damaged Gandhi's prestige in the country since it was believed that he had so much influence over the Viceroy that he could pursuade the latter to spare the lives of these young revolutionaries.

The execution of Bhagat Singh gave a rude shock to all, especially to the youths of this land, who held him in highest esteem. This supreme sacrifice and contribution of these young revolutionaries is the obvious to talk about here. Their self-sacrifice, courage, fearlessness, patriotism had thrilled the hearts of young and old alike in the country. The official historian of Congress Pattabhai Sitaramayya observed that at the moment Bhagat Singh's name was as popular as Gandhi. The former, according to Netaji Subhas Chandra Bose, "was the symbol of spirit of revolt which has taken possession of the country from one end to the other." Similarly, Pandit Jawaharlal Nehru declared at the Karachi Session of the Congress that in Bhagat Singh self-sacrifice and bravery had passed the upper limits.

The Mahatma himself had the special favour for the brave patriots. Subhas Chandra and a host of other Congressmen, criticise Gandhi for the Gandhi-Irwin Pact. But he gave the following assurances to the Congressmen at Bombay:

(1) He would ask the Karachi Congress for a mandate to bind the hands of the Congress deputationist to the Round Table Conference;

(2) That mandate would contain nothing that was not consistent with the status of Independence for which Lahore Congress had declared; and

(3) He would use all his influence and strain every nerve to secure amnesty for those who had been left out in the Pact.

When Gandhi was travelling from Bombay to Delhi, he got the news that the government had decided to execute Bhagat Singh and two of his comrades in the Lahore Conspiracy Case. Pressure was brought to bear upon Gandhi to intercede with the Viceroy for the commutation of their death sentence. 'On this occasion', writes Subhas Chandra Bose, "I ventured the suggestion that he should, if necessary, break with the Viceroy on this question because the execution was against the spirit, if not the letter of the Delhi Pact. I was reminded of a similar incident during the armistice between the Sinn Fein Party and the British government when the strong attitude adopted by the former had secured the release of an Irish political prisoner sentenced to the gallows". But the Mahatma who did not want to identify himself with the revolutionary prisoners would not go far and it naturally made a difference when the Viceroy realised that the Mahatma would not break on that question. However, at that time, Lord Irwin told Gandhi that he had received a largely signed petition asking for the commutation of the death sentence passed on the three Lahore prisoners. He would postpone their execution for the time being and give serious consideration to the matter, but beyond that he did not want to be pressed at the moment. The conclusion which Gandhi and everybody else drew from the attitude of the Viceroy was that the execution would be finally cancelled and there was a jubilation all over the country. A few even gave credence to a rumour which, of course, was denied by Gandhi's Secretary, that the Mahatma was in correspondence with Bhagat Singh in that connection.

The general expectation being that the execution would be cancelled, it was a painful and expected surprise when on March 24, Gandhi received the news that Bhagat Singh and his comrades had been hanged the night before.

Was Gandhi cornered by the Viceroy? Was Gandhi cornered by the victroy or it was the failure of his techniques is a moot question. But there is no doubt that he made sincere efforts to save those persons' precious lives in spite of his disapproval of their reliance on violence for the liberation of their motherland. He pleaded with the Viceroy to the maximum of his ability to commute their death sentence on the ground that commutation was the general demand of the country and, therefore, any act to the contrary would endanger internal peace. Giving an idea of the efforts that he made in this connection, he wrote in *Young India,* "I pleaded with the Viceroy as best as I could. I brought all the persuasion at my command to bear on him. On the day fixed for the final interview with Bhagat Singh's relatives, I wrote a personal letter to the Viceroy on the morning of 23rd. I poured my soul in it but of no avail".[19]

Even the Viceroy himself admitted that Gandhi tried his best to save Bhagat Singh and his colleagues. He told Gandhi that there was a very minor difference between a revolutionary and a terrorist. The same person can be a high profile revolutionary and a patriot for one section of society and a terrorist and a criminal for the government. This is the difference of the perception. The Viceroy even strictly told Gandhi that why he being a votary of non-violence was pleading for the violent persons. The Mahatma told him that the question at present was not between violence and non-violence but of saving the precious lives. The Viceroy gave only a verbal assurance.

Almost three decades after the event when Lord Irwin became Viscount Halifax, he wrote the following in his autobiography: Mr. Gandhi said he greatly feared that unless I could do something, the effect would be to destroy our Pact. I said, I would regret this no less than he, but would be clear to him. There were only three possible causes. "The *First* was to do nothing and let the execution proceed. The *Second* was to change the order and grant Bhagat Singh reprieve, the *Third* was to hold up any decision till after the Congress meeting was over". Gandhi requested him to show mercy and spare the young men. He appealed to the Viceroy's conscience.

Gandhi was blamed for their execution. He was greeted with hostile slogans and black flags when he went to Karachi to attend the session of the Congress. He was blamed for not making commutation of the sentence as a condition for the settlement between the Congress and the Government in the Gandhi-Irwin Pact. Bose laid emphasis on it while presiding over the session of *Naujawan Bharat Sabha* at Karachi on March 21, 1931. Perhaps Bose was aware of the intentions of the government and even before this meeting, he pleaded for making commutation of the sentence as a pre-condition for the Pact.

Gandhi admitted that he could not go to that extent since India's Freedom for him was more important than the execution of a few youngmen who would cheerfully court death for the freedom of their country. The Congress Working Committee endorsed this policy Gandhi explained in *Young India*, "I might have made the commutation a term of the settlement. It could not be so made and to threaten withdrawal would be a breach of faith. The Working Committee had agreed with me in not making commutation a condition precedent to truce".[20]

Nevertheless, Gandhi paid his tribute to the courage and spirit of self-sacrifice of Bhagat Singh and his comrades while drafting a resolution on their sacrifices for the Karachi Session: "This Congress, while disassociating itself from and disapproving of political violence in any shape and form, places on record its admiration of the bravery and sacrifice of the late Sardar Bhagat Singh and his comrades Sukhdev and Rajguru and mourns with bereaved families the loss of these lives. The Congress is of the opinion that the Government has lost the golden opportunity of promoting goodwill between the two nations admittedly held to be essential at this juncture and of winning over to the method of peace the party which being driven to despair to political violence".

Sardar Kishan Singh, the father of late Sardar Bhagat Singh came to the rostrum of the Karachi Congress and spoke in favour of the resolution. Though Gandhi drafted this resolution, he never wanted anyone to emulate the act of violence committed by the young revolutionaries. He only admired the spirit behind their action. But such subtle distinctions could hardly fall to the notice of the masses and so Gandhi had to issue a statement shortly after Karachi Session, repudiating the cult of violence that came to be associated with the name of Bhagat Singh. He wrote in *Young India*, "Bhagat Singh's character about which I had heard so much from reliable sources carried me away and identified me with the cautions and balanced resolution passed at Karachi. I regret to observe that the caution has been thrown to the winds, the deed itself being worshipped as it is worthy of emulation."

D.P. Das accused Gandhi of a small deception for a cause which he considered noble and patriotic. Das concludes, "Gandhi's deception should not make us cynical about him nor should the discovery of his deception for a cause considered noble by him. He was a great Indian. Let Bhagat Singh episode be just a spot in his career and a condonable spot. Have we not condoned the spots in the moon, 'Yudhisthira and Rama'?" Das is unduly harsh on Gandhi, as from all available accounts Gandhi did his best.

It must, however, be remembered that Gandhi was not the Viceroy of India. All that he could do was to try. And he pleaded hard. The Viceroy on the other hand was not an old Mughal ruler. He was a paid employee of the British Government. He must have acted under the directions. Gandhi, on the other hand, worked hard for the settlement and was determined not to break. In fact, his earnest desire was to make absolutely final what was provisional.

Gandhi's role has, therefore, to be judged in view of the fact that the ultimate goal for which he stood for was the Country's Independence, where not one but many Bhagat Singh's were required to be sacrificed for the sake of freedom. "It has", said Gandhi, "increased our power for winning freedom for which Bhagat Singh and his comrades have died".

REFERENCES

1. Bakshi, S.R., *The Struggle for Independence, op. cit.*, pp. 59-60.
2. Nehru, Jawaharlal, *An Autobiography* (Delhi: Oxford Press), 1999, pp. 175-176.
3. Sandhu, Virendra, *Amar Shahid Bhagat Singh, op. cit.*, p. 46.
4. Gopal, S., *Jawaharlal Nehru, A Biography*, Volume 1, 1889-1947 (London: Oxford University Press), pp. 394-95.
5. Home Department File Number 4/12/1931.
6. *Ibid.*
7. *Ibid.*
8. Home, Political, File Number 4/10/1930.
9. *Ibid.*
10. *Ibid.*, File Number 4/20/1931.
11. *Hindustan Times*, February 19, 1931.
12. Home Department, Judicial, File Number 152/1/1931.

13. *Ibid.*
14. *The Collected Works of Mahatma Gandhi*, Volume XLV (New Delhi: Publications Division, Ministry of Information and Broadcasting, Government of India), p. 200.
15. *Young India*, March 12, 1931.
16. *Leader*, March 12, 1931
17. Deol, G.S., *Shaheed-e-Azam Sardar Bhagat Singh, op. cit.*, pp. 86-87.
18. *The Collected Works of Mahatma Gandhi*, Volume XIV, *op. cit.*, pp. 315-316.
19. *Young India*, 02-04-1931.
20. *Ibid.*

BIBLIOGRAPHY

Adhikari, G.D. (ed.), *Documents of the History of the Communist Party of India*, 3 Vols. (New Delhi: People's Publishing House), 1971-82.

Adorno, Theodore (ed.), *The Authoritarian Personality* (New York, Harper and Row), 1950.

Aiyar, S.P. (ed.), *The Politics of Mass Violence in India* (Bombay: P.C. Manaktala & Sons), 1967.

Ambedkar, B.R., *What Congress and Gandhi have done to the Untouchables* (Bombay: Thacker & Co.), 1946.

Argov, Daniel, *Moderates and Extremists in the Indian National Movement, 1883-1923,* Bombay, 1968.

Ayer, S.A., *Unto Him a Witness* (Bombay: Thacker & Co.), 1951.

Ayer, S.A., *Selected Speeches of Subhas Chandra Bose* (New Delhi: Publications Division, Ministry of Information and Broadcasting, Government of India), September 1974.

Azad, Baba Prithivi Singh, *The Legendary Crusader* (Bombay: Bharatiya Vidya Bhavan), 1987.

Azad, Maulana Abul Kalam, *India Wins Freedom* (Madras: Orient Longman), 1989.

Ba Maw, *Breakthrough in Burma, Memoirs of a Revolution, 1939-1946* (London: Yale University Press), 1968.

Bakshi, S.R., *Bhagat Singh—Patriot Martyr* (New Delhi: Anmol Publications), 1990.

Bakshi, S.R., *The Struggle for Independence* (New Delhi: Anmol Publications), 1989.

Basu, Aparna, *The Growth of Education and Political Development in India, 1898-1920* (Delhi: Oxford University Press), 1974.

Bell Charles, *Indian Mutiny* (London: Clarendon Press), 1931.

Bhattacharya, Buddhadeva (ed.), *Freedom Struggle and Anushilan Samiti* (Calcutta: Anushilan Samity), 1979.

Bimal Prasad (ed.), *A Revolutionary Quest* (Delhi: Oxford University Press), 1980.

Bose, Arun Coomar, *Indian Revolutionaries Abroad, 1905-1922* (Patna: Bharati Bhavan), 1971.

Bose, Nirmal Kumar, *Studies in Gandhism* (Calcutta: Merit), 1962.

Bose, Nirmal, *My Days with Gandhi* (Calcutta: A. Chatterjee), 1953.

Bose, Sarat Chandra, *I Warned My Countrymen* (Calcutta: Netaji Research Bureau), 1968.

Bose, Subhas Chandra, *An Indian Pilgrim: An Unfinished Autobiography and Collected Letters* (Bombay: Asia Publishing House), 1965.

Bose, Subhas Chandra, *Dissentient Report* (Kodalia: S.C. Bose), 1961.

Bose Subhas Chandra, *Impression in Life* (Lahore: Hero Publication), 1947.

Bose, Subhas Chandra, *The Indian Struggle* (Calcutta: Published for Netaji Publishing Society), 1948.

Bose, S.C., *Samagra Rachanabali*, Vol. 1 (Calcutta: Ananda Publishers), 1980.

Bose, Sisir K., *Remembering My Father: A Centenary Offering* (Calcutta: Netaji Research Bureau), 1988.

Brass, Paul R., *Language Religion and Politics in North India* (Delhi: Vikas), 1975.

Brown, Judith M. (ed.), *Eminent Indians* (Calcutta: Shanti Mitra), 1946.

Burham, James, *The Managerial Revolution* (England: Penguin Books Ltd.), 1942.

Chaman Lal, *Martyrs of India* (Kanpur: Chaman Lal), 1957.

Chandra, Bipan, *Communalism in India* (New Delhi: Vani Educational Books), 1984.

Chandra, Bipan, *India's Struggle for Independence 1857-1947* (New Delhi: Viking), 1998.

Chandra, Bipan, *Modern India* (New Delhi: NCERT), 1976.

Chandra, Bipan, *Nationalism and Colonialism in Modern India* (New Delhi: Orient Longman), 1979.

Carr, E.H., *International Relations Between the Two World Wars* (London: MacMillan & Co. Ltd.), 1952.

Cashman, Richard, *The Myth of the Lokamanya Tilak and Mass Politics in Maharashtra* (Berkeley: University of California Press), 1975.

Chatterji, Dilip Kumar, *C.R. Das and Indian National Movement* (Calcutta: Postgraduate Book Mart), 1965.

Chatterjee, Santimony (ed.), *Collected Works of Meghnath Saha* (Bengali) (Calcutta: Asia Publishing House), 1993.

Chattopadhyaya, Sarat Chandra, *Sharat Sahitya Samagra,* Vol. 2 (Calcutta: Anand Publisher), 1986.

Chaudhary, Sukhbir, *Growth of Nationalism in India,* 2 Vols. (New Delhi: Trimurti Publications), 1973.

Chirol, Valentine, *The Indian Unrest* (New Delhi: Lights and Life Publications), 1979.

Chirol, Valentine, *India Old and New* (New Delhi: Lights and Life Publications), 1975.

Chitkara, M.G., *Converts do not Make a Nation* (New Delhi: APH Publishing Company), 1998.

Collins, Larry and Lapierre, Dominique, *Freedom at Midnight* (New Delhi: Vikas Publishing House), 1997.

Chopra, P.N. (ed.), *Quit India Movement: British Secret Report* (Faridabad: Thomson Press), 1976.

Dalton, Dennis, *Indian Idea of Freedom* (New Delhi: Manhohar), 1982.

Das, Cittaranjan, *India for Indians* (Madras: S. Ganesh), 1921.

Das, Durga, *India from Curzon to Nehru* (London: Collins), 1959.

Das, Harihar, *Subhas Chandra Bose and the Indian National Movement* (New Delhi), 1983.

Dasgupta, Hemendranath, *Deshbandhu Chittaranjan Das* (Delhi: Government of India, Publications Division), 1960.

Datta, K.K., *Renaissance, Nationalism and Social Changes in Modern India* (Calcutta: Bookland), 1965.

Datta, V. N., *Madan Lal Dhingra and the Revolutionary Movement* (New Delhi: Vikas Publishing House), 1978.

Dawarkadas, Kanji, *India's Fight for Freedom: An Eye Witness Story, 1913-37* (Bombay: Popular Prakashan), 1960.

Desai, A.R., *Social Background of Indian Nationalism* (Bombay: Popular Prakashan), 1976.

Deol, G.B., *Shaheed Bhagat Singh* (Patiala: Publications Bureau, Punjabi University), 1985.

Deol, Gurudev Singh, *Shaheed-e-Azam Sardar Bhagat Singh* (Ludhiana: Deep Prakashan), 1978.

Dharamvira, *Bhai Parmanand aur Unka Yug* (New Delhi: Bhai Parmanand Smarak Samiti), 1981.

Dharamvira, *Lala Hardayal and Revolutionary Movements of His Times* (New Delhi: Indian Book Company), 1970.

Dignon, Don, *The Indian Revolutionary Problem in British Diplomacy 1914-1919* (New Delhi: Allied Publishers), 1983.

Dwivedi, Surendranath, *The Quest for Socialism* (New Delhi: Radiant), 1984.

Elsbree, Willard H., *Japan's Role in Southeast Asian Nationalist Movements, 1940 to 1945* (New York: Russel and Russel), 1953.

Fawcett, C.B., *Frontiers—A Study in Political Geography* (London: OUP), 1921.

Gandhi, M.K., *An Autobiography or The Story of My Experiments With Truth* (Ahmedabad: Navajivan Publishing House), 1983.

Gandhi, M.K., *Hind Swaraj* (Ahmedabad: Navajivan Publishing House), 1983.

Ganpule, N.G., *Netaji in Germany* (Bombay: Bhartiya Vidya Bhavan), 1959.

Ghose, Sankar, *Political Ideas and Movements in India* (Calcutta: Allied Publishers), 1975.

Ghosh, Ajoy, *Bhagat Singh and His Comrades* (Bombay: Allied Publishers), 1945.

Ghosh Ajoy, "Bhagat Singh and His Comrades" in New Age, 22 March 1931.

Ghosh, Chitra, "*The Socio-Political Ideas of Subhas Chandra Bose*", (Unpublished Paper, Presented in the Seminar of J.N.U. on 10-11 November, 1997).

Ghosh, Kalicharan, *The Roll of Honours: Ancedotes of Indian Martyrs* (Calcutta: Vidya Bharati), 1975.

Gooch, G. P., *Nationalism* (London: Longman), 1945.

Gopal, Madan, *Life and Times of Subhas Chandra Bose* (New Delhi: Vikas Publishing House Pvt. Ltd.), 1978.

Gopal, S., *Jawahar Lal Nehru, A Biography*, Volume 1, 1889-1947 (London: Oxford University Press).

Gordon, Leonard A., *Bengal: The Nationalist Movement, 1876-1940* (New York: Columbia University Press), 1974.

Gordon, Leonard, A., *Brothers Against the Raj* (Delhi: Viking), 1990.

Grover, Verinder, *Subhas Chandra Bose* (Delhi: Deep & Deep), 1999.

Guha, A.C. (ed.), *The Story of Indian Revolution* (Calcutta: Asia Publishing House), 1972.

Guha, A.C., *First Spark of Revolution* (Delhi: Orient Longman), 1972.

Guha, R. (ed.), *Subaltern Studies III* (Delhi: OUP), 1989.

Gupta, A.C. (ed.), *Studies in the Bengal Renaissance, in Commemoration of the Birth of Bipin Chandra Pal* (Calcutta: Asia Publishing House), 1958.

Gupta, Manmath Nath, *Bhagat Singh and His Times* (Delhi: Lipi Prakashan), 1977.

Gupta, Manmath Nath, *They Lived Dangerously: Reminiscences of a Revolutionary* (Delhi: Peoples Publishing House), 1969.

Gupta, Manmath Nath, *History of the Indian Revolutionary Movement* (Delhi: Somiya Publications Pvt. Ltd.), 1969.

Gupta Manmohan, *Ek Krantikari Ke Sansmaran* [Allahabad: Indian Press (Publications) Pvt. Ltd.], 1969.

Gupta, Vishwa, Prakash, and Mohini Gupta, *Mahatma Gandhi: Vyakti aur Vichar* (Hindi) (New.Delhi: Radha Publications), 1995.

Gupta, Vishwa Prakash and Mohini Gupta, *Subhas Chandra Bose: Vyakti aur Vichar*, (Hindi) (New Delhi: Radha Publications), 1997.

Gupta, V.P. and Mohini Gupta, *The Quest for Freedom—A Study of Netaji Subhas Chandra Bose* (New Delhi: Radha Publications), 1998.

Gustave, Le Bon, *The Psychology of Revolution* (London: Adelphi Terrace Leipsic Inselstrasse, 20, English Translation), 1913.

Gwyer, Sir Maurice and Appadorai, A., *Speeches and Documents on the Indian Constitution 1921-1947*, 2 Vols. (Delhi: Oxford University Press), 1957.

Haithcox, John P., *Communism and Nationalism in India: M.N. Roy and Comintern Policy, 1920-1939* (Princeton: Princeton University Press), 1971.

Haniz L. and Others, *The Individual Psychology of Alfred Adler* (New York, Harper and Row), 1967.

Harbick, Walter and Alexander Werth, *Netaji in Germany* (Calcutta: Netaji Research Bureau), 1970.

Hardayal, *Hints for Self Culture* (Delhi: Rajkamal Publications), 1948.

Hasan, Mushirul, *Nationalism and Communal Politics in India 1916-1928* (New Delhi: Manohar Publishers), 1981.

Hasan, Mushirul, *Communal Pan-Islamic Trends in Colonial India 1916-1928* (New Delhi: Manohar Publishers), 1981.

Hayashida, Tatsuo, *Netaji Subhas Chandra Bose* (Bombay: Allied), 1970.

Hitler, Adolph, *Mein Kampf* (Boston: Houghton Mifflin), 1943.

Holmes, T.R., *History of Indian Mutiny* (London: Macmillan), 1898.

Hugh, Toya, *Springing Tiger: A Study of Subhas Chandra Bose* (Calcutta: International Netaji Seminar), 1976.

Iyer, N. *Raghavan, Moral and Political Writings of Mahatma Gandhi* (Oxford: Clarendon Press), 1986, Vols. 1 - 3.

Jinnah, M.A., *Speeches and Writings* (Lahore: Mustaffa), 1960.

Jog, N.C., *Lokamanya Bal Gangadhar Tilak* (New Delhi: Publications Division, Ministry of Information and Broadcasting, Government of India), 1990.

Josh, Sohan Singh, *Hindustan Ghadar Party—A Short History* (New Delhi: People Publishing House), 1977.

Kapoor, A.N. and V.R Gupta, *A Dictionary of Gandhian Thought* (Delhi: Ambe Books), 1995.

Karan Singh, *Prophet of Indian Nationalism* (Bombay: Bharatiya Vidya Bhawan), 1991.

Karunakaran, K.P. (ed.), *Modern Indian Political Thought* (New Delhi; Allied Publishers), 1962.

Kaushik, P.D., *The Congress Ideology and Programme, 1920-47* (New Delhi: Allied), 1964.

Ker, James Campbell, *Political Trouble in India* (Calcutta: Superintendent Government Printing), 1917.

Khan, Shahnawaz, *My Memories of INA and Its Netaji* (Delhi: Rajkamal Publication), 1946.

Khan, Shah Nawaz, *I.N.A. and Its Netaji* (Delhi: Allied Publishers), 1946.

Khatri, Ramakrishna, *Shaheedon Ki Chaya Mein* (Nagpur: Vishwa Bharati Prakashan), 1983.

Khullar, K.K., *Shaheed Bhagat Singh* (New Delhi: Hem Publishers Pvt. Ltd.), 1981.

Khushwant Singh and Satindra Singh, *Ghadr, 1915* (New Delhi: Penguin), 1966.

Laski, Harold, *Grammar of Politics* (London: Allen), 1948.

Lebra, Joyce C., *Jungle Alliance: Japan and the Indian National Army* (Singapore: Asia Pacific Press), 1959.

Lohia, Rammanohar, Guilty Men of India's Partition (Allahabad: Kitabistan), 1960.

Maikap, S.C., Challenge to the Empire: A Study of Netaji (New Delhi: Publications Division, Government of India), 1995.

Majumdar, R.C., *History of the Freedom Movement in India* (Calcutta: Firma K.L. Mukhopadhyaya), 1963.

Malleson, G.B. and Kye John, *History of Indian Mutiny* (London: Allen), 1879.

Masani, R.P., *Dadabhai Naoroji* (New Delhi: Publications Division, Ministry of Information and Broadcasting, Government of India), 1960.

Masselos, Jim, *Indian Nationalism: An History* (New Delhi: Sterling), 1985.

Mathur, L.P., *Indian Revolutionary Movement in USA* (Delhi: S. Chand), 1970.

Mazumdar, R.C., Roy Chaudhary, H.C. Dutta, Kali Kumar, *An Advanced History of India* (Madras: Macmillan), 1950.

Mazumdar, S.N., *In Search of a Revolutionary Ideology* (New Delhi: People Publishing House), 1979.

McCarthy, Justin, *History of Our Times*, Vol. II (London: Chatts), 1897.

Mead, Henery, *The Sepoy Revolt: Its Causes and Consequences* (Delhi: Gyan), 1986.

Menon, V.P., *The Transfer of Power in India* (Bombay: Orient Longman), 1957

Mishra, B.B., *The Indian Political Parties* (New Delhi: Oxford University Press), 1999.

Mody, Homi, *Pherozeshah Mehta* (New Delhi: Publications Division, Ministry of Information and Broadcasting, Government of India), 1963.

Mohan, Kamlesh, *Militant Nationalism in the Punjab, 1919-35* (New Delhi: Manohar Publications), 1985.

Mukherjee, Radhakamal, *India: The Dawn of New Era* (New Delhi: Radha Publications), 1997.

Nanda, B.R., *Gandhi, Pan-Islamism, Imperialism and Nationalism in India* (Bombay: Oxford University Press), 1989.

Nath, Shaileshwar, *Terrorism in India* (New Delhi: National Publishing House), 1980.

Nayar, Kuldip, *The Martyr: Bhagat Singh Experiments in Revolution* (New Delhi: Har-Anand), 2000.

Nehru, Jawaharlal, *An Autobiography* (London: The Bodley Head), 1958.

Noorani, A.G., *The Trial of Bhagat Singh: Politics of Justice* (New Delhi: Konark Publishers Pvt. Ltd.), 1996.

Pandit, V.L., *The Scope of Happiness* (New Delhi: Vikas Publishing House), 1965.

Parmanand, Bhai, *The Story of My Life* (New Delhi: S. Chand), 1982.

Parekh, Bhikhu, *Colonialism, Tradition and Reform* (Delhi: Sage Publications), 1999.

Philips, C.H. (ed.), *The Evolution of India and Pakistan 1858-1947 Select Documents* (London: George Allen & Unwin), 1962.

Planning Commission, *Subhas Chandra Bose: Pioneer of Indian Planning* (New Delhi: Planning Commission, Government of India), 1997.

Prasad Bimal, "Remembering Bhagat Singh" in *Mainstream*, 28 March 1981.

Radhakrishnan, S. (ed.), *Mahatma Gandhi* (London: George Allen), 1949.

Rai, Lala Lajpat, *The Story of My Life* (New Delhi: D.A.V. Publications), 1987.

Rai, Lala Lajpat, *Young India: An Interpretation and a History of the Nationalist Movement from Within* (New Delhi: Publications Division, Ministry of Information and Broadcasting, Government of India), 1968.

Raj, Sukhdeo, *Jab Jyoti Jagi* (Mirzapore) 1971.

Rajendra Prasad, *An Autobiography* (Bombay: Asia Publishing House), 1957.

Ralhan, O.P. (ed.), *Hindu Mahasabha*, 2 Vols. (New Delhi: Anmol), 1997.

Ray, Sibnarayan (ed.), *M.N. Roy, Philosopher and Revolutionary* (Delhi: Oxford University Press), 1987.

Reddy, P.L. Sanjeev and Mishra, S.N. (eds.), *Sardar Vallabhbhai Patel: The Nation Builder* (New Delhi: Indian Institute of Public Administration), 2004.

Report of the Sedition Committee (Calcutta: Superintendent Government Printing Press), 1918.

Rolland Romain, *The Life of Vivekananda and the Universal Gospel* (Calcutta: Advaita Ashrama), 1953.

Rolland, Romain, *Mahatma Gandhi*, translation from French by Catherine D. Groth (New Delhi: Publications Division, Government of India).

Romain Rolland and Gandhi Correspondence (New Delhi: Publications Division, Government of India), 1976.

Roy, Dilip Kumar, *The Subhash I Knew* (Bombay: Asia Publishing House), 1946.

Saggi, P.D. (ed.), *Life and Works of Netaji Subhas Chandra Bose* (Bombay: Overseas Publishing House), n.d.

Saigal, Omesh, *Shaheed Bhagat Singh, Unique Martyr in Freedom Movement* (New Delhi: Gyan Publishing House), 1976.

Sandhu, Virendra, *Shaheed Bhagat Singh* (Delhi: Anmol Pulishers), 1974.

Sarkar, J. and others (eds.), *Bhagat Singh and His Times in India's Freedom Struggle* (New Delhi: People's Publishing House), 1986.

Sarkar, M.L. (ed.), *Subhas Bose: 1939-40* (Bengali), Translate by Nipen Chakravarty (Calcutta: Forward Publishing Concern), 1975.

Sarkar, S., *Modern India: 1885-1947* (Delhi: Macmillan), 1983.

Savarkar, V.D., *Six Glorious Epics in Indian History* (New Delhi: Rajdhani Granthagar), 1970.

Savarkar, V.D., *Swantantryaveer Savarkar: A Guide to Indian Revolutionary Movement* (Pune: Maharashtra Prantik Hindu Sabha), 1963.

Savarkar, V.D., *The Indian War of Independence 1847* (New Delhi: Phoenix), 1947.

Savarkar, V.D., *The Story of My Transportation of Life* (Bombay: Sadbhakti Publications), 1950.

Savarkar, V.D., *Echo from Andamans* (Poona: Venus Books), 1947.

Savarkar, S.S. & Joshi G.M., *Historic Statements of V.D. Savarkar, V.D* (Bombay: Popular Prakashan), 1967.

Seale, Patrick and Maureen McConville, *Philby* (London: Penguin Books Ltd.), 1978.

Sen, Rajen (ed.), *Deshbandhu Chitta Ranjan* (Calcutta: Ananda Publishers), 1927.

Shankar, Ramahari, *Gandhi's Encounter with Indian Revolutionaries* (New Delhi: Siddharth Publications), 1996.

Shankhdher, M.M., *Reflections on Indian Politics* (New Delhi: Kumar Brothers), 1973.

Sharma, Jai Narain, *Power, Politics & Corruption: A Gandhian Solution* (New Delhi: Deep and Deep), 2004.

Sharma, Malikarjuna, *Role of Revolutionaries in the Freedom Struggle* (Hyderabad: Marxist Study Forum), 1987.

Shraddhanand, Swami, *Inside Congress* (Bombay: Phoenix), 1946.

Singh, Gulab, *Under The Shadow of Gallows: A Story of a* Revolutionary (Delhi: Rup Chand), 1964.

Singh, Swarn, *Path of Revolution: A Bibliography of Shaheed* Bhagat Singh (Dehli: Anmol Publications), 1990.

Singhal, Jwala Prasad, *Sphinx Speaks: The Story of Pre-historic Nations* (New Delhi: Sadgya), 1963.

Sitaramayya, Pattabhi, *History of Indian National Congress* (Delhi: S. Chand & Co.), 1969.

Speeches by Lord Irwin (Shimla: Government of India Press), 1931, Vol. 2.

Sri Aurobindo, *Sri Aurobindo on Himself* (Pondicherry: Sri Aurobindo Ashram), 1972.

Srivastava, Harindra, *Five Stormy Years: Savarkar in London* (New Delhi: Allied Publishers), 1983.

Suda, J.P., *Main Currents of Social and Political Thought in India,* Vol. III (Meerut: K. Nath & Co.), 1973.

Sullivan, O. Noel (ed.), *Terrorism, Ideology and Revolution* (Sussex: The Harvester Press), 1986.

Swami Jyotiramayananda (ed.), *Vivekananda, with a Garland of Tributes* (Madras: S. Ganeshan), 1988.

Swami Nikhilananda, *Vivekananda* (Calcutta; Advaita Ashram), 1999.

Tara Chand, *History of Freedom Movement in India* (New Delhi: Publications Division, Ministry of Information and Broadcasting, Government of India), 1962.

Tendulkar, D.G., *Mahatma*, 8 Vols. (New Delhi: Publications Division, Ministry of Information and Broadcasting, Government of India), 1958.

Thakur, Gopal, *Bhagat Singh, The Man and His Ideas* (New Delhi: Anmol Publishers), 1989.

The Collected Works of Mahatma Gandhi, 100 Volumes (New Delhi: Publications Division, Ministry of Information and Broadcasting, Government of India), 1958-1995.

Toye, Hugh, *Subhash Chandra Bose: The Springing Tiger* (Delhi: Jaico Publishing House), 2001.

Transcript of an Interview with Jaidev Gupta, Nehru Memorial Museum and Library, New Delhi.

Transcript of an Interview with Chabil Das, Nehru Memorial Museum and Library, New Delhi.

Transcript of an Interview with Durga Das Khanna, Nehru Memorial Museum and Library, New Delhi.

Varma, V.P., *Modern Indian Political Thought* (Agra: Luxmi Narain Agarwal), 1961.

Vajpeyi, J.N., *The Extremist Movement in India* (Allahabad: Chug Publications), 1974.

Wardlow, Grant, *Political Terrorism Theory, Tactics and Counter Measures* (Cambridge: Cambridge University Press), 1982.

Walport, Stanley, *Jinnah of Pakistan* (New Delhi: Oxford University Press), 1965.

Walport, Stanley, *Tilak & Gokhale* (Bombay: University of California Press), 1962.

Wilkinson, Paul, *Political Terrorism* (New York: Macmillan), 1976.

Wasti, Syed Razi, *Lord Minto and the Indian National Movement 1905-1910* (London: Oxford University Press), 1965.

Yadav, K.C. and Akiko Sen (eds.), *Subhas Chandra Bose* (New Delhi: Hope India), 2003.

Zangwill, Israel, *Principles of Nationalities* (London: Heinemann), 1920.

Index